EARTH QUAKE SURVIVAL MANUAL

Lael Morgan

Epicenter Press
Fairbanks/Seattle

Earthquakes: A Global View

The black dots on this map represent seismic hot spots throughout the world.

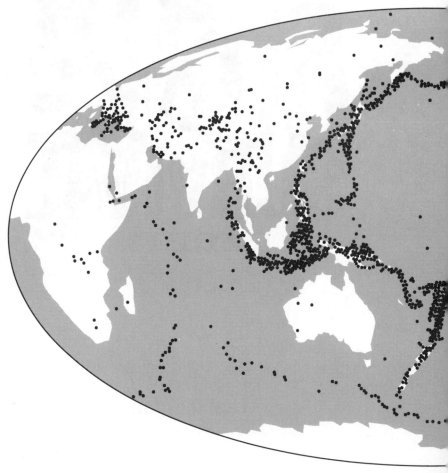

EARTH QUAKE SURVIVAL

MANUAL

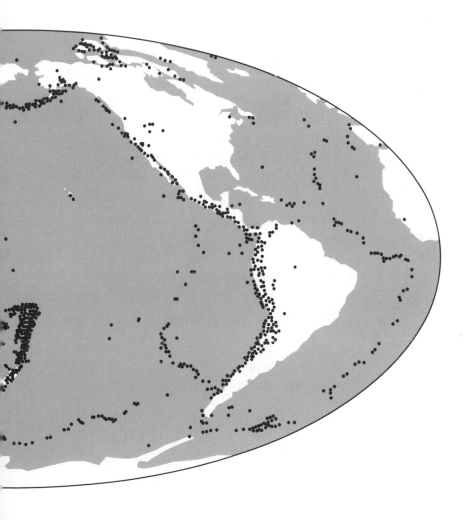

Book design: Newman Design/Illustration
Cover: Leslie Newman
Printer: McNaughton & Gunn
Editor: Don Graydon
Proofreader: Jean Andrews
Info-graphics: Ben Garrison
Maps: Susan K. Goter, U.S. Geological Survey/National Earthquake
Information Center
Text © 1993 Lael Morgan

Library of Congress Cataloging-in-Publication Data

Morgan, Lael.
    Earthquake survival manual / by Lael Morgan.
        p.   cm.
    Includes bibliographical references and index.
    ISBN 0-945397-20-8 (pbk.) : $14.95
    1. Earthquakes—Safety measures.   2. Survival skills.   I. Title.
QE539.2.S34M67   1993
363.3'495—dc20   93-27804
                                              CIP

To order single copies of the EARTHQUAKE SURVIVAL MANUAL, send
$14.95 (Washington state residents add $1.23 sales tax), plus $5 for priority-
mail shipping, to: Epicenter Press, 18821 64th Ave. NE, Seattle, WA 98155, or
call (206) 485-6822 to place an order by phone.
Booksellers: Retail discounts are available from our trade distributor,
Graphic Arts Center Publishing Co. of Portland, Ore. Phone 800-452-3032.

PRINTED IN THE UNITED STATES OF AMERICA
First printing, October, 1993

Contents

Illustrations

Charts & Informational Graphics

Maps

Quick Earthquake Guide

WHAT TO DO RIGHT NOW

❏ Locate the safe spots and the dangerous spots where you live and work.

❏ Develop an earthquake action plan with family, neighbors, co-workers, and school authorities.

❏ Decide with family and friends on a contact point *outside* your quake zone where you can check in if local communication lines are down.

❏ Know how, when, and where to shut down utilities at their source.

❏ If you live in an old house, see if it needs reinforcing or bolting to the foundation.

❏ Secure heavy appliances like water heaters. Anchor heavy furniture, bookshelves, and computers. Cushion good china and crystal. Install quake-proof latches on cupboards.

❏ Store volatile and flammable liquids outside.

❏ Prepare emergency supplies—food, water, first aid kit, and cash—to last at least seventy-two hours. Store somewhere easily accessible to your family.

❏ Keep a jackknife and flashlight with your personal gear.

❏ Maintain a battery-operated radio.

❏ Don't let your automobile gas tank run low.

❏ Back up business and family records to maintain outside your fault zone.

❏ Order a copy of *The Earthquake Survival Guide* for a friend or loved one (from Epicenter Press, 18821 64th Avenue NE, Seattle, WA 98155; 206-485-6822).

Quick Earthquake Guide

WHAT TO DO DURING THE QUAKE

❏ Don't panic! Duck, cover, and hang on!

❏ Although the majority of earthquakes can be measured in seconds, it is possible to survive a much longer roller like the big one in Anchorage, Alaska, in 1964, which lasted several minutes.

Indoors:

❏ Don't dash for the exit unless you are clearly imperiled. Stairs can buckle. Elevators can jam. Buildings may shed windows and outer walls. Panicking crowds may be more dangerous than the quake.

❏ Dive under a heavy table or desk or into a small room or hallway that does not have overhead fixtures or much ceiling area to support. Take shelter in a doorway only if it shows no sign of buckling and has no swinging door to mash your fingers.

❏ If in bed, stay there and cover up unless you are clearly imperiled.

❏ Avoid windows and glass dividers, hanging planters, bookcases, office chairs with casters, chimneys, kitchens, and shopping mall walkways.

Outdoors:

❏ Stay clear of building exteriors (especially brick and glass), trees, phone and power lines, gas and water mains, fuel tanks, dams, reservoirs, and possible avalanche areas.

❏ If on the waterfront, head immediately for high ground.

❏ If driving, get your vehicle as far off the road as possible in an area clear of bridges and underpasses, and stay with it unless you are clearly in danger. Public parking garages and double-decked freeways are particularly vulnerable to quakes.

Anywhere:

❏ Do not use candles or any open flame because of possible gas leaks.

❏ Prepare mentally and physically for aftershocks.

The Quick Earthquake Guide has been designed for ease of photocopying. Permission is granted for private, non-profit duplication of these pages only for free distribution to friends, relatives and neighbors. Copies of the full EARTHQUAKE SURVIVAL MANUAL may be ordered by sending $14.95 (Washington state residents add $1.23 sales tax), plus $5 for priority-mail shipping, to: Epicenter Press, 18821 64th Ave. NE, Seattle, WA 98155.

Quick Earthquake Guide

WHAT TO DO AFTER THE QUAKE

❏ Wearing shoes is a must. Beware of broken glass.

❏ Do not move anyone who is seriously injured if not necessary. You may find hospitals badly damaged or even abandoned.

❏ Use caution in rescue missions. Buildings may be undermined.

❏ Shut off power, gas, and water at source only if you suspect damage.

❏ If you suspect damage to utility lines and wiring, do not turn utilities back on without expert help.

❏ Turn on your radio for information on danger areas, corridors of safe travel, and aftershocks.

❏ Your toilet tank and water heater may be needed as emergency water sources. Don't drain them.

❑ Phones off the hook can shut down the system. Use sparingly.

❑ Think carefully before endangering yourself to search for loved ones. If your planning has been good, chances are they are safe.

❑ If you venture out, leave a note telling where you are headed.

❑ Be wary of both looters and military troops sent to protect you from looters. Both may shoot.

❑ Don't trust unknown water sources. Boil or chlorinate.

❑ Remember that survivors, especially children and household pets, may be in shock and will require careful reassurance. You have nothing to gain by hysterics.

Dedication

Dedicated to Catherine Breslin, who monitored the East Coast while I kept track of the West, and to Jacq Bachmann Siracusa, who shares my enthusiasm for seismology to a fault. Also to my publisher, Kent Sturgis, who was so fascinated with this subject that he named our company Epicenter Press and would not rest until this book saw print.

Preface

When I was a girl, an earthquake shook our home in rural Maine.
"Make it stop, Mommy!" pleaded my five-year-old brother. Our mother, who was nearly six feet tall, an expert horsewoman, and prone to moving things like our living room couch all by herself, admitted that stopping an earthquake was beyond her, and we were unnerved. Gone forever was our belief that adults were invulnerable, and with it our childish trust in the "solid earth" beneath our feet.

My practical response was to develop a taste for living on the edge. Today I own homes in both Alaska and California, the number one and number two most earthquake-prone states in the USA.

My actual experience with quakes is spotty. When the giant earthquake hit my home base in Alaska in 1964, I was watching waterspouts from the deck of a 36-foot-schooner in the Caribbean. Anchorage, which had been built slapdash as a railroad boomtown after the turn of the century, was rebuilt to become the All-American City it might never have otherwise been.

A bit grumpy for having missed the excitement, I resettled in Fairbanks, Alaska, in time to experience triple quakes there (5.4, 5.6, and 5.4 on the Richter scale, on June 21, 1967). Then on to California and the big time.

On the morning of February 9, 1971, the San Fernando Valley quake tossed me rudely out of bed in Laguna Beach, California. I was more than 100 miles from the epicenter of the quake that killed sixty-four people and injured 2,400. Damage was minimal in my area, but the shock was frightening enough so that I never again grumbled about missing a seismic event.

Nevertheless, two years later I purchased my house in San Juan Capistrano, right behind the rubble of a historic Catholic mission downed by a quake from the San Andreas Fault in 1812, taking forty fervently praying parishioners off to their just reward. In 1985 the Fathers built a replica, reinforcing the original design with structural steel, but I still worry about it. Are the church fathers ignoring something God was trying to tell us in 1812?

I began research for this book in 1981. Having rented out my Capistrano house, I was enjoying a hiatus in Los Angeles with a friend who owned a delightful mansion on an earthquake fault in the shadow of the famous Hollywood sign. Halfway into my research I learned that despite the luxury of my host's mansion, I was living in the most hazardous of all structures—an old, unreinforced brick showplace from the 1920s, perched on a steep hillside and directly below a similar structure poised above us on slender pilings. I talked to my host about reinforcing his structure, but the job was too expensive to be practical, he told me. The odds have favored him, to date.

In 1987 I was filling in as head of University Relations at Cal State Northridge when a quake from the Whittier Fault shut the school down. The campus—36,000 students—was evacuated, but I had to stay on to answer telephones that wouldn't connect to anyone across town but kept ringing in call from reporters from New York. Our most noticeable loss was a chandelier that never should have been installed in a fault zone. But at our sister school fifty miles away, Cal State Los Angeles, a student was killed by a falling concrete slab.

Shortly thereafter I moved back to Alaska, where I purchased a second home. I bought earthquake insurance. And I was careful to cultivate the correct mental attitude of a fault-zone resident: never become emotionally attached to a house. I also decided I'd rather weather a quake here than on the East Coast, where the concept of earthquake preparedness ranks right up there with planning for an invasion from Mars.

My favorite T-shirt reads, "Sooner or later, everything east of the San Andreas Fault is going to plummet into the Atlantic." The theory has scientific merit, because both Boston and New York have formidable faults. Most Easterners chuckle at the idea, but I wish they'd give it at least half as much thought as those of us who have come to gauge our lives by the Richter scale.

Chapter 1
AT RISK

While the United States experiences only 2 percent of the world's earthquakes, some 90 percent of its population lives in seismically active zones. Canada's most quake-prone area is sparsely settled British Columbia, but two major cities—Vancouver and Victoria—are clearly at risk there.

If this worries you, there are a few "safe" places to choose from. For example, portions of Texas, Mississippi, Alabama, and Florida appear to be seismically safe—but these states have a dreadful track record with tornados and hurricanes, so it's a matter of picking your poison.

Most vulnerable to seismic disturbances are Alaska, the least populated state, and California, where densely settled areas are threatened. Twin quakes that ripped across California's Mojave Desert in the summer of 1992 caused scientists to reconsider the odds for a "really big one" there in the near future.

Also listed as major earthquake risk areas by the Federal Emergency Management Agency are Portland, Oregon; Seattle; Salt Lake City; Memphis, Tennessee; and Charleston, South Carolina.

Vancouver, Seattle, and Portland also are in the seismic spotlight in view of recently discovered evidence of a large uplift and sudden subsidence in the sea floor along the Pacific Northwest coast, along which runs an undersea fault that some call the most ominous in North America. James C. Savage and Michael Lisowski, geologists with the U.S. Geological Survey at Menlo Park, California, found that this fault is slowly squeezing together the mountains along the Olympic Peninsula in Washington while the coastline rises, clear signs of major earthquake stress.

Even more thought-provoking is the finding of terrestrial plant roots more than 310 feet (roughly 95 meters) below sea level on the continental shelf in a fault area about 12½ miles (about 20 kilometers) north of Vancouver Island. According to Dr. John Luternauer of the Geological Survey of Canada, radiocarbon dating showed the roots to be 10,500 years old. The core sample that contained them also showed pollen indicating that lodgepole pine, alder, hemlock, grasses, ferns, and sphagnum moss once grew on what is now ocean floor.

Savage and Lisowski believe the fault off the Pacific Northwest coast shows potential to spawn

A Union Pacific Railway track was damaged when hillside fill gave way near Olympia, Wash., in an April, 1965, earthquake.

a quake measuring a magnitude up to 9 that could last about three minutes, cause coastal areas to drop as much as six feet, and generate huge waves from Vancouver to Mendocino, California. Such a shaker would destroy most of the buildings in Portland, they say.

Most residents of Oregon, Washington, and British Columbia have been complacent because no great earthquake has occurred in this area since European settlers first arrived. However, new findings indicate that the coast in this region experienced one or more massive earthquakes about 300 years ago and that giant rockers may strike in a 400- to 500-year sequence.

Planners in Vancouver spent millions to reinforce bridges and strengthen water and sewer lines to withstand quakes up to 8 on the Richter scale, but the city is not set to ride out a magnitude of 9. Washington state, which has weathered rattlers of up to 7.4 and annually commemorates a 1949 quake that took eight lives, continues to consider tighter building codes, as does Portland, which is built on loose ground. But standard construction restrictions in this region generally have been no stricter than those of upstate New York or New England.

Not that the East Coast can afford to be complacent. New England weathered such a violent series of temblors in the 1700s that Puritan preachers enjoyed a wholesale conversion of sinners for fear of them. "At his Wrath the Earth shall tremble," they preached from Jeremiah 10:10 as the earth shook.

There are some formidable faults under Boston, and New York City may be even more imperiled. In the 1980s, Dr. Robert Ketter, a professor of seismology at the State University of New York, predicted that a "major, destructive earthquake" would hit the east within the next twenty years. Ketter's warning is echoed by Ian Buckle, deputy director of the National Center for Earthquake Engineering Research, in Buffalo, New York, who believes the East Coast can expect an earthquake more devastating than the Loma Prieta quake that struck California in 1989, killing sixty-seven people and causing an estimated $6 billion in damage. Another seismic expert, professor Peter H. Mattson of Queens College, New York City, claims there are enough fault lines along 125th Street in Manhattan, under Long Island Sound, and in upstate New York near Indian Point to take the threat seriously. Flushing Meadows and the land beneath Kennedy International Airport, areas composed of weak soil, could be particularly hard hit.

Unfortunately, "seismic loading" is a term foreign to New York building codes, said television anchorman Chuck Scarborough, who put together a three-part series on New York quake hazards in 1988. He called the situation "terrifying." "Consider the architecture," Scarborough said. "Consider the water mains, steam pipes, gas mains that would all fracture and disgorge their contents."

Until recently, residents of southcentral states showed as little concern for quake dangers as easterners, despite the fact that some of the strongest shocks in the recorded history of North America occurred in New Madrid, Missouri, in 1811 and 1812. Three jolts, ranging to a maximum of about 8 on the Richter scale, were reported to have briefly reversed the course of the Mississippi River and rattled buildings as far away as Boston and Chicago. There were few casualties because the country was sparsely settled.

Today this area has a large population and, although there has been little seismic activity since the New Madrid quake, residents are ahead of most in earthquake preparedness because of a recent false alarm. In early 1990, self-proclaimed climatologist Iben Browning predicted a 50-50 chance for a major quake along the new Madrid Fault about December 3. Although he lacked credentials to make such a pronouncement, Browning had forecast California's Loma Prieta quake within a week. Citizens of southeastern Missouri and neighboring areas of Arkansas and Tennessee took him seriously enough to pack survival seminars, conduct earthquake drills, and stockpile water and food. New Madrid even closed its schools December 3-4, and there were few hard feelings when disaster evaded them.

"I think Iben Browning Day is here to stay," Jim Bradley, chairman of the New Madrid County Fault Commission, told The Associated Press one quake-free year later. "He did more in just a few words to heighten our awareness of a serious problem than anybody, and he'll never be forgotten."

Ironically, Anchorage, Alaska, scene of the biggest earthquake ever recorded in North America—8.5 on the Richter scale, on Good Friday, 1964—may not be as well prepared for another big one as New Madrid, even though scientists say the area is ripe for another hit.

Death toll from the Good Friday earthquake was only 11, with another 120 people from Alaska to Oregon killed from ensuing tsunamis (tidal waves). Since that time, the government has es-

Historic Quakes Worldwide

Selected earthquakes

Date	Place	Magnitude*	Deaths**
1556: Jan. 24	Shaanxi Province, China	Unknown	830,000
1755: Nov. 1	Lisbon, Portugal	Unknown	30,000 – 100,000
1811-1812: Three quakes	New Madrid, Missouri	Maximum about 8.0	Several
1906: April 18	San Francisco, California	8.25	700
1908: Dec. 28	Messina, Italy	7.5	85,000 – 160,000
1923: Sept. 1	Tokyo/Yokohama, Japan	8.3	100,000 – 150,000
1946: June 23	Vancouver Island, Canada	7.2	1
1960: May 21-30	Chile	8.5	5,000
1964: March 27	Alaska	8.5	131
1965: April 29	Seattle, Washington	6.5	7
1971: Feb. 9	San Fernando, California	6.5	64
1976: July 28	Tangshan, China	7.8	More than 242,000
1985: Sept. 19	Mexico City, Mexico	8.1	More than 9,000
1988: Dec. 7	Armenia	6.9	25,000
1989: Oct. 17	Loma Prieta, California	7.1	67
1990: June 20	Iran	7.7	More than 40,000
1992: June 28	Mojave Desert, California	7.4	1

* Authorities differ to some degree on the magnitudes determined for particular earthquakes. Magnitudes shown are based on the Richter scale or a closely related scale.

** Authorities differ widely on the number of deaths from particular earthquakes.

Source: Government and commercial publications

tablished the Tsunami Warning Center at Palmer, Alaska. However, it is not faith in the warning system but a particular mental attitude that accounts for limited concern here, according to Robert Page, of the USGS in Menlo Park.

"In Alaska, we have a situation where a lot of people who survived the 1964 earthquake don't feel that earthquakes are that much of a problem because they survived that very big one," Page said. "But a lot has changed since 1964. Anchorage is a lot more vulnerable."

In 1964 the population of Anchorage was a mere 50,000; it's now 250,000. People today depend more on public services and if a quake caused power and gas-line failures when temperatures were twenty to thirty degrees below zero, it could be especially tragic. Many structures built in Anchorage since 1964 are multi-storied, often on stilts above ground-level parking. "That sort of structure didn't fare well in San Francisco during the 1989 earthquake," seismologist Page notes.

In contrast to Alaska, California launched an intensive preparedness program in the 1980s and continues to demand earthquake readiness of its citizens as the state's population grows. When talk spread that the 1989 Loma Prieta quake, with a magnitude of 7.1, had lowered odds for another seismic event in the San Francisco area, USGS officials quickly countered with stern warnings.

"First, Loma Prieta was not the big one. It was a moderately big one, certainly destructive to some parts of the Bay Area, but nowhere near the size of the great San Francisco earthquake of 1906," spokesmen wrote in the widely distributed booklet titled *The Next Big Earthquake In The Bay Area May Come Sooner Than You Think.*

"Second, having an earthquake like Loma Prieta has little to do with the likelihood of having another one on a different fault, somewhere else in the area."

Scientists, however, debunk the well-worn movie cliche that the state will eventually disappear into the sea in a cataclysmic shake-up. "There is nothing in California's geologic history which suggests massive subsidence, so there is no reason to predict it in the immediate future," notes USGS scientist Tom Burdett.

"Even if western California is one day an island in the mid-Pacific, it will still be above sea level," echoes geologist Haroun Tazieff.

California was the scene of North America's second-biggest earthquake, the 1906 San Francisco temblor (magnitude 8.25) that killed more than 700 people. The state has suffered many quakes registering over 7 on the Richter scale and numerous smaller ones that also caused considerable damage. But compared with other quake-prone parts of the world, California's record does not look so bad. One of the biggest earthquakes ever recorded struck a 500-mile stretch of central Chile in May of 1960 with a Richter magnitude of 8.5, releasing many times the energy of the Loma

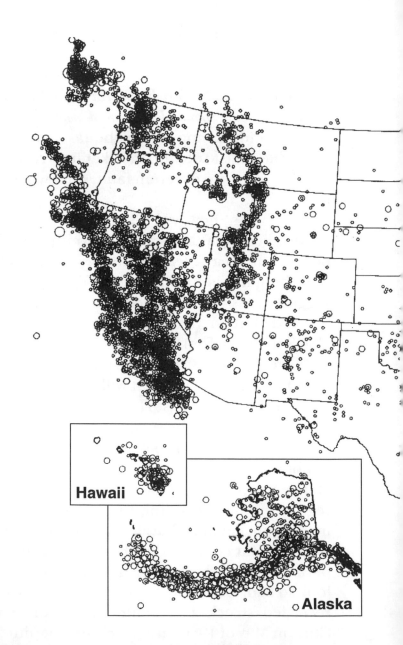

Hawaii

Alaska

U.S. Geological Survey/National Earthquake Information Center

Prepared by Susan K. Goter

Seismic Activity in the U.S.

From 1899 to 1990

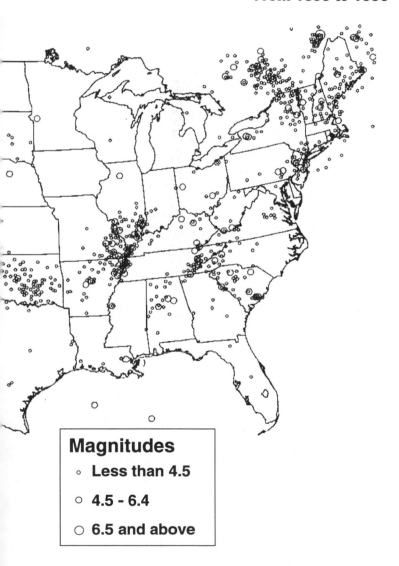

Magnitudes

- ∘ **Less than 4.5**
- ○ **4.5 - 6.4**
- ◯ **6.5 and above**

Prieta quake and resulting in a death toll of more than 2,000.

A quake in Shaanxi Province, China, in 1556 took 830,000 lives, the largest number of deaths ever recorded in a seismic event. Tangshan in the Hebei Province of northern China sustained a hit of about 7.8 in 1976, with an official death toll of 242,000 (unofficially about 750,000).

A 1923 quake of 8.3 in Tokyo and Yokohama lasted just 76 seconds but ranks as the greatest natural disaster of modern times in Japan. Loss of water and a cyclonic fire storm (*taksumaki*) added to the disaster, which cost more than 100,000 lives. Quakes in Lisbon, Portugal (1755), and Messina, Italy (1908), each killed tens of thousands of people.

The extraordinary death tolls of early quakes and of those in the third world in modern times can be blamed in large part on primitive building codes and densely populated areas. Because America is a younger nation, even its older buildings make a better showing. And modern structures designed for earthquake country have already proved their worth.

San Francisco, America's most earthquake-conscious city, has shored up its old landmarks, going so far as to mandate removal of gargoyles and ornamental drain pipes that were popular during the Victorian era. The Loma Prieta earthquake subjected the city to a severe jolt, but only a relatively small number of buildings were lost.

Earthquake Risk in the U.S.

Seismic-risk map based on horizontal acceleration

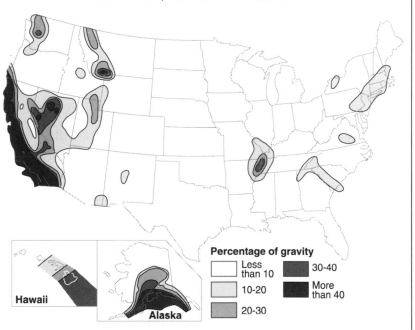

Percentage of gravity

- Less than 10
- 10-20
- 20-30
- 30-40
- More than 40

Hawaii

Alaska

Acceleration during ground movement is a critical factor in the design of earthquake-resistant buildings. Numbers refer to the maximum horizontal acceleration during an earthquake (as a percentage of the acceleration due to gravity). The probability that an acceleration of the amount indicated will occur in any given 50-year period is 1 chance in 10.

Source: U.S. Geological Survey

"On the whole, the city came through it remarkably well, using emergency plans efficiently and coping with unforeseen obstacles heroically," *Newsweek* magazine reported. "The experience also revealed points of vulnerability and secret pockets of strength—in the process teaching important lessons of survival not only for San Francisco but other cities facing sudden catastrophe."

Some people complain about footing ever-larger relief costs for others who court disaster by crowding into earthquake-prone areas like the Bay Area and who build homes on beachfront, landfill, and fault lines.

"You could move 80 percent of California's population to South Dakota or Nebraska and virtually eliminate the hazard," observed James Wright, an expert on natural disasters, following Loma Prieta. "But people want to live in California."

And, fortunately, with what we're learning about earthquakes, people who invest in earthquake planning have a good chance of surviving even on shaky ground.

Chapter 2
CAUSES & PREDICTIONS

According to a legend of Siberia's Kamchatka Peninsula, the cause of earthquakes was the flea-bitten dog team of a god named Tuli, who pulled the earth around on his sled. When the dogs stopped to scratch, quakes ensued, with Kamchatka getting decidedly more than its share.

Other cultures also had explanations for sharp motions of the earth. Mayans saw our globe as a cube supported at the corners by four gods who kept an eye on overpopulation, shaking off surplus weight when the birth rate went up. Early Algonquin Indians thought their land rode on the back of a giant tortoise. Three pillars—faith, hope, and charity—were believed to support early Romanian realms; when one pillar faltered, the world tottered for want of balance.

From the beginning, earthquakes disrupted lives, but even the more sophisticated societies relied on mystical explanations. When beautiful Catholic Lisbon, the much-envied port city of Portugal, was devastated in 1755 by a quake, with the loss of as many as 100,000 lives, Anglican churchmen smugly called it God's rightful punishment for a rich, wicked, and superstitious city.

Yankee preacher Thomas Paine let his imagination soar in a sermon in which he declared an earth- quake that terrified his parishioners at Weymouth, Massachusetts, to be "an Alarm of God given to New England... which as a Thief in the Night surprised us the evening after the Lord's Day, Oct. 29, 1727."

Developing an earthquake science

One of the first to look beyond the wrath of the divine in studying quakes was a Chinese man named Chang Heng, who in about A.D. 132 invented a magnificent device designed to indicate earthquake direction. The elaborately decorated copper vessel, eight feet in diameter, displayed eight dragon heads near its base, each dragon holding a ball. The force of a quake was supposed to cause one of the dragons to drop its ball into the open mouth of a frog figure mounted below, indicating the direction of the quake.

The first careful documentation of a quake was the work of priests following the Lisbon disaster. About a decade later, Sir William Hamilton, British envoy to the court of Naples, took an interest in the Vesuvius volcano, which was active at that time. Hamilton investigated a 1783 quake in Calabria, Italy, for the Royal Society and theorized on the connection between quakes and volcanic eruptions, something scientists still argue about.

Robert Mallet, an Irish engineer, performed experiments in the 1850s leading to seismic wave calculations. Luigi Palmieri, director of the Vesuvius volcano observatory during this same period, developed the first seismograph, complete with clock and recording drum.

John Milne, a British subject who taught geology and mining at the Imperial College of Engineering in Tokyo, proposed creation of a Seismological Society of Japan following a quake near Yokohama in 1880. By the time Milne left the country twenty years later, Japan had established 700 observation stations. Nineteenth-century earthquake research grew into a new field called seismology (from a Greek word meaning trembling earth).

An earthquake is sudden rapid shaking of the earth, most often caused by the breaking and shifting of rock beneath the surface. For this to make much sense, you must also understand that the earth's rocky surface layer is divided into huge sections called plates, which are in slow but constant motion.

By 1912, German geophysicist Alfred Wegener had collected evidence showing that the continents had once been joined together. His work was largely ignored until the early 1960s, when Fred Vine and Drummond Matthews discovered that the ocean floor was spreading apart at midocean ridges. By 1968, this evidence had been worked into the full-blown theory of plate tectonics, based on observations that the earth's plates—perhaps afloat on molten rock and now thought to number about twelve—seem to be jockeying for position, with some sliding under or over others.

Areas in which plates interact are aptly called faults, and the risk of earthquake increases when two plates snag. Pressure builds when plates lock, until the rocks cannot hold any longer and they shift, causing the earth to tremble. Most large earthquakes occur along these plate boundaries, but they can also occur elsewhere. The most famous exceptions to plate boundary earthquakes in the United States were the New Madrid, Missouri, quakes of 1811-1812.

Earthquakes also occur in conjunction with volcanic activity and landslides. Man-made explosions, especially of nuclear origin, have also produced quakes.

In 1902, Italian Giuseppe Mercalli worked out a scale to rank earthquakes in terms of their intensity at any given location, based on eyewitness reports of the impact on people and property.

The Richter Scale

An increase in magnitude of 1 unit on the Richter scale equals an increase of about 30 times in the amount of energy released. A magnitude 3 earthquake releases 30 times as much energy as a magnitude 2 quake, and so on up the scale.

Richter magnitude	Relative energy release
1	1
2	30
3	900
4	27,000
5	810,000
6	24,300,000
7	729,000,000
8	21,870,000,000
9	656,100,000,000

In 1935, American Charles Richter developed a system to measure the magnitude of a quake, which is its actual size independent of how intensely it might be felt at a particular spot. Seismograph readings are used in determining the Richter magnitude.

Since earthquakes come in a vast range of sizes, Richter designed his scale in a way that avoids the unwieldy numbers that would result if the scale simply started at 1 and went on up into the millions. On the Richter scale, a magnitude 2 earthquake releases about thirty times the energy of a magnitude 1 quake. Likewise, a magnitude 3 earthquake releases about thirty times the energy of a magnitude 2 quake, and so on up the scale. The very largest earthquakes release billions of times as much energy as the barely felt magnitude 3 shocks.

Predictions

Scientists have long watched faults—especially the San Andreas Fault, a dramatic 800-mile-long gash that runs within thirty miles of Los Angeles and within one mile of San Francisco—with the hope of making earthquake predictions. But it hasn't been easy. In 1983, a quake that registered 6.5 on the Richter scale nearly destroyed Coalinga, California, in an area where no active fault was mapped. In 1987, a similar quake under Whittier, California, caused seismologists to investigate previously unsuspected fault networks deep underground, representing a whole new class of hazards.

Since ancient times, man has attempted earthquake prediction with decidedly mixed results. In the 1600s, Capuchin Fathers at Melfi, Italy, discovered that foaming water in a nearby lake provided timely warnings of earthquakes. It was noted in the 1700s that falling well-levels preceded quakes in Portugal, Spain, and Italy. Successful predictions have been reported based on strange, pale rainbows appearing a few hours before a tremor; on "earthquake weather" (sullen and sultry); on wispy clouds, vapor trails, and mysterious night lights. But unsuccessful predictions based on the same phenomena far outnumber those that score.

Observers have noted odd behavior and restlessness in animals prior to earthquakes: screaming pheasants, jumpy zoo residents, wild animals vacating the danger area. The Japanese spent nearly $1 million in researching the catfish as a quake indicator, though that experiment was terminated as unsuccessful in 1992.

Other pre-quake phenomena have been reported: emissions of radon gas, changes in electromagnetic fields, particular juxtapositions of the earth and sun and moon, and variations in sound waves bouncing off irregularities around a fault. Based on the prediction of Chinese scientists studying a variety of precursors, the city of Haicheng, Liaoning Province, was evacuated just before a quake wrecked 90 percent of that city in 1975. However, they failed to predict the Tangshan quake of 1976 which claimed more than a quarter of a million lives (and at the same time they made the embarrassing mistake of evacuating another city where no quake struck).

Statisticians ransacking old records have found no clear pattern of when earthquakes strike. They occur at irregular intervals in all seasons, at night as well as day.

In 1985, a group of seismologists announced America's first large-scale earthquake prediction experiment. Noting enormous strain on a section of the San Andreas Fault near Parkfield (almost halfway between Los Angeles and San Francisco), they surrounded the site with seismic instruments. Their best guess for the occurrence of a major quake was January 1988, but they hedged by declaring a window that extended until "sometime before 1993." As if in mocking response, the quake-prone area calmed down. Noting a decline in even minor activity in 1986, scientists moved their prediction date to March 1991, but the time passed without incident.

Scientists continue to try to predict earthquakes based on calculations of accumulated strain in fault-prone areas. In 1988, the National Earthquake Prediction Evaluation Council announced a 50 percent probability of an earthquake of magnitude 7 or larger in the San Francisco Bay Area within thirty years. Following the 7.1 Loma

Prieta quake of the following year, the official guess became a 67 percent chance of another earthquake of the same size or larger in the Bay Area within the following thirty years.

U.S. Geological Survey

Without warning, a three-story structure in San Francisco's Marina District collapsed in the 1989 Loma Prieta quake.

A 7.4-magnitude quake that slammed though the sparsely populated Mojave desert on June 28, 1992, may have illustrated the current limitations of scientific earthquake prediction. The quake left seismologists astonished and shaken, according to a report in the *New York Times* by reporter Sandra Blakeslee: "Astonished because this earthquake did weird things like setting off earthquakes more than 500 miles to the north, ripping across four existing faults to make a single larger fault… Shaken because the nation's most dangerous fault, the San Andreas, just got more dangerous."

The Mojave seismic event, which apparently touched off a second quake of 6.5 magnitude and dozens of high-powered aftershocks, pointed up the danger of a "really big quake" in the near future. "Everyone agrees that Los Angeles and San Bernardino should treat this like a final warning," said Dr. Allan Lindh, director of the U.S. Geological Survey's seismology branch in Menlo Park, California. "It's like when you clean up camp. It's time to make that last pass through our cities, homes, and lives and act as if the darned thing will happen tomorrow."

Chapter 3

PREPARE NOW!

How do you prepare for something that might occur in the next two seconds or in 300 years?

The first step is to get it straight in your own head that survival is possible. To further justify the effort, realize that earthquake preparedness will also prove useful in disasters like floods, tornados, winter storms, evacuation due to a spill of hazardous materials, failure of atomic power plants, and civil disorders. It's something you should do anyway!

Psychologically prepare

It's very important to realize up front that you may not have the help of any government agency for seventy-two hours. (That's the time line given by Los Angeles city planners, who ask citizens to plan accordingly). Water, electricity, phones, and other utilities considered basic by most urban dwellers may be down for a lot longer, so self-sufficiency is your goal.

Assure family and loved ones that their chances for survival are excellent. Then prepare them for the possibility that you may not be able to get to them for several days—and maybe for as long as two weeks. Warn them not to put too much stock in rumors. (During the Long Beach, California, quake of 1933, it was widely believed that Catalina Island had disappeared. In 1964, as Alaska reeled from aftershocks of the biggest quake ever recorded in North America, a hysterical broadcast reporter told listeners that Anchorage was "a sea of flames," which was equally ridiculous).

Now make your disaster planning so sound that you, your family, friends, and co-workers will all have faith in it.

Communicate

Careful planning now will help you maintain communication with your family even during a major disaster.

Choose a primary and secondary location for family members to meet after the quake. (However, make it clear that no one should attempt to travel unless it is safe. More people may be killed or injured trying to get to loved ones who are safe than by the quake itself.)

Check with your emergency planning agencies to learn of hazards (like underground gas

Here is a sample of an emergency planning card useful for family members:

Family Disaster Plan

Name of family member _____

Blood type _____ Date of birth _____

Home Address _____

Home Phone _____

Emergency
meeting place _____ Phone _____
                   (In your neighborhood)

Emergency
meeting place _____ Phone _____
                   (Outside your neighborhood)

Address of
family contact _____
                   (Outside your fault area)

Day phone _____ Evening phone _____

lines) on accustomed routes and of designated safe corridors of transport. Discuss alternative routes with your family. Learn what sites will be designated as evacuation centers and homeless shelters.

Choose someone out of your fault area with whom you and other members of your family can check in by phone. Rescue workers generally set up portable phones and will allow you one call. You may be able to phone out of state even when disaster damage prevents calls across town. Make sure your children know how to call long distance.

Issue every family member an identification card on which his or her blood type is noted. Add to that the meeting locations you've chosen and the check-in phone number. (See the sample family disaster planning card in this chapter.) Post emergency numbers by your phone.

Make special plans for infants, the elderly, and handicapped people. Make arrangements with neighbors to look out for each other's families and pets if necessary. Check with school officials on policy for dismissing children in an emergency.

It's wise to review your plans every six months or so, and quiz your children to be certain they remember what to do.

Around the House

Precautions you can take to help make your home safer in an earthquake

Replace heavy hangings over bed with lightweight alternatives

Nail plywood to ceiling joists to protect occupants from falling chimney bricks

Anchor hanging lamps with closed hooks or relocate

Keep heavy, unstable objects away from exit routes and anchor wheels

Secure latches on cupboards to prevent doors from swinging open

Secure top-heavy furniture to wall studs with metal braces

Keep breakables in low or secure cabinets

Keep fire extinguisher in accessible place

Locate main electrical and gas switches for emergency shut-off

Keep emergency supplies on hand, including drinking water, canned or dried foods, first-aid kit, flashlight, and portable radio with extra batteries

Use flexible connectors where gas lines meet appliances

Stabilize water heater with metal straps to wall studs

Federal Emergency Management Agency/American Red Cross

Shore up for shock

You can begin getting your house, school, or workplace into good order by starting with a walk through the building, discussing safe and unsafe areas with others who use the facility. Draw up a floor plan showing two exits if the building covers an extensive area.

Make sure everyone in your group knows how to turn off utilities at the source, has the tools to do so, and practices doing it. Make it clear, though, that utilities should remain on unless damaged, particularly important in the case of natural gas because too many shutdowns at one time could rupture the main.

If your building is not structurally sound, inves-tigate strengthening it (see Chapter 6, on "Safer Building"). Pay particular attention to homes not bolted to their foundations, unreinforced masonry, large glass frontage, and inadequate roof bracing.

Secure your water heater with straps bolted to the wall. Be sure the water heater and other gas appliances are connected to the gas supply with a flexible line. Anchor other heavy appliances, furniture, file cabinets, and computers. Invest in childproof latches (that prevent cupboard doors from opening more than an inch without extra action) or in other secure latches. Cushion glassware, secure shelves, and anchor whatnots to shelves with Velcro. Store heavy items on lower shelves.

Secure Your Water Heater

Use straps bolted to studs inside the wall. Be sure the water heater and other gas appliances are connected to the gas source with a flexible line to reduce the danger of the line breaking during an earthquake.

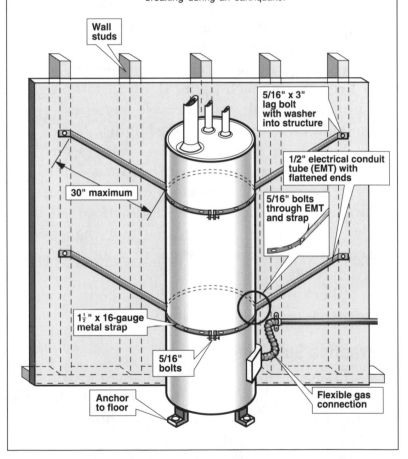

Wall studs

5/16" x 3" lag bolt with washer into structure

1/2" electrical conduit tube (EMT) with flattened ends

30" maximum

5/16" bolts through EMT and strap

$1\frac{1}{2}$" x 16-gauge metal strap

5/16" bolts

Anchor to floor

Flexible gas connection

Avoid putting plants in hanging pots, which can become lethal in a good shake. Remember that chandeliers don't do well in earthquake country. Move beds away from hanging mirrors, pictures, and windows, and make sure bed headboards are secure. Have blinds and curtains over windows while you are sleeping so that shattering glass will fall outward. Store a collapsible ladder or fire rope on each upper floor. Install plug-in night lights (with rechargeable batteries) that can be used as emergency flashlights.

Move hazardous materials—cleaners, fertilizers, chemicals, and petroleum products—outside your living area, if possible. Put them in sturdy containers fastened to the wall or floor. At the very least, station them near the garage door. Keep your fire extinguishers in good working order, ready for action.

Check on the cost of earthquake insurance (see Chapter 7, "Earthquake Business").

Practice Surviving

Practice "duck, cover, and hang-on" drills by locating and ducking under a piece of heavy, protective furniture. Get a good hold on the furniture so that you can travel with it in case it moves during a quake. Be sure everyone understands how to use your fire extinguishers. Take a first-aid class, and encourage friends, family, co-workers, and neighbors to join you. Conduct evacuation drills at your home and business. Practice locating and shutting off utility valves.

Duck, Cover and Hang On

When the shaking starts, duck under a solid, sheltering piece of furniture like a table or a desk. Be sure your head and face are covered to protect them from broken glass and falling objects. Hang on to the piece of furniture, and if it moves, hang on and go with it.

Stay supplied

Preparedness is not complete without laying in the supplies you need for self-sufficiency during an emergency. The Red Cross and other disaster preparedness agencies offer advice on what supplies to stock. Following are lists of items to store at home, at work, in your car, and to carry with you. You can tailor these lists to your family's needs.

Store foods and other supplies where everyone in your group can get to them. In the home, a handy closet with bearing walls and little ceiling to topple in is usually a good choice.

Store at home

Basic supplies

- ☐ Flashlights, spare bulbs, and batteries with ten-year shelf life
- ☐ Lantern
- ☐ Candles
- ☐ Matches
- ☐ Flares
- ☐ Radio with extra batteries
- ☐ Wrench for gas turn-off
- ☐ Well-maintained fire extinguishers with ABC rating
- ☐ Garden hose for siphoning and fire fighting
- ☐ Plastic sheeting and staple gun for window replacement

- Dust mask to trap airborne bacteria
- Saw
- Ax
- Crow bar
- Shovel
- Whistle
- Rope
- Duct tape
- Protective glasses
- Gasoline in safety cans
- Mace for protection
- Fire resistant strongbox with cash, family papers, and photos (for tracing missing people and for emotional comfort)
- Portable toilet, toilet paper, and garbage bags for waste; portable-toilet chemicals available at RV supply stores (chlorinated lime will also serve)
- Prescription drugs to last for two weeks
- Feminine hygiene supplies
- Infant supplies, if necessary
- Spare eyeglasses (the drugstore variety should serve)
- First-aid kit
- Soap
- Compact cook stove
- Cooking equipment
- Can opener

❑ Canteen

❑ Water purification kit (see section in this chapter on water)

❑ Heat source for cold country (might double as a cook stove)

❑ Sleeping bags or blankets

❑ Heavy shoes or boots

❑ Extra clothing

❑ Tent

❑ Knapsack

❑ Duffel bag

❑ Stationery and stamps

❑ City map

❑ This book

❑ A good medical book, such as the *Merck Manual of Diagnosis and Therapy* (Merck, Sharp & Dohme Research Laboratories, 1992).

❑ Boy Scout Fieldbook or a survival guide (see Sources section at back of this book)

Some useful luxuries:

❑ Portable generator

❑ Bicycle

❑ Books and games for children

Store some of these supplies in an evacuation kit—a duffel bag or knapsack that can be carried by one person. Included should be your tent, compact cook stove, cooking gear, sleeping bags, canteen, water purification kit, and other basics.

Food

Store canned juice and ready-to-eat food for your family and pets to last at least seventy-two hours. Check periodically to make sure it has not outlasted its shelf life and that cans are not bulging or broken.

Use within six months:

❑ Evaporated milk

❑ Dried fruit in metal container

❑ Dry crisp crackers in metal container

Use within a year:

❑ Nonfat dry or whole milk in metal container

❑ Canned meat, poultry, fish, fruit; mixtures of meat, vegetables, or cereal products in sealed cans or jars

❑ Canned condensed meat soups and vegetable soups; dehydrated soups in metal containers

May be stored indefinitely:

❑ Sugar

❑ Salt

❑ Cereals ready to eat in metal container or uncooked

❑ Hydrogenated fats, vegetable oils

❑ Sweets and nuts, hard candy, instant pudding

❑ Coffee, tea, instant cocoa, dry cream products, bouillon products, flavored extracts, baking soda, baking powder

Water

Store a case of bottled water in quarts for carrying with you if evacuation becomes necessary. Use five-gallon containers for home use. Estimates vary on how much water is needed per person for drinking and cooking. In moderate weather, half-a-gallon per day should serve, while a gallon per day is recommended in hot climates. If storage space is available, fill a fifty-five-gallon plastic drum halfway with a garden hose, add one ounce of liquid chlorine bleach to purify the water, and fill to the top. Empty the drum every six months and refill with fresh water and chlorine.

Other water sources to be considered in an emergency are toilet tanks (not bowls), the water heater, and ice cubes. Water from swimming pools can be used for washing but not for drinking. (The chlorine won't hurt you, but salts and acids used in pool maintenance can cause kidney damage).

Never trust a water source after a quake, even your city water supply. You can purify water by boiling it five to ten minutes or by using commercial preparations of halazone or Globaline following directions on the package.

Household bleach containing hypochlorite (preferably 5.25 percent) may be used, according to the table below. Let the water stand for thirty minutes before using:

Amount of water	Clear water	Cloudy water
1 quart	2 drops	4 drops
1 gallon	8 drops	16 drops
5 gallons	½ tsp.	1 tsp.

Household tincture of iodine may also be used in the same manner as bleach.

Store at your workplace

- High-energy food bars
- Comfortable walking shoes
- Whistle
- Flashlight and batteries
- Transistor radio

Consider the possibility of theft and store items inconspicuously. Set aside your rattiest running shoes, and hide some of your food supplies in the toes.

Store in your car

- Canteen with water
- Emergency food supply
- Camping gear
- Sleeping bag
- Knapsack
- Sturdy shoes or boots
- First-aid kit
- Critical medication
- Transistor radio
- Fire extinguisher
- Hose for siphoning
- Tool kit
- Crow bar
- Flashlight
- Flares
- Matches

❑ Light sticks

❑ Paper and pencil

❑ Change for pay phone

❑ Compass

❑ City and state maps

❑ Cold-weather gear for northern regions

❑ Sun protection for southern regions

Get the best city map available, like the Thomas Guide that shows alternative routes, freeway overpasses, bridges, dams, reservoirs, and hospitals. Check out frequently driven routes, noting freeway overpasses, bridges, and other potential highway dangers that could force you to reroute or abandon your vehicle during an emergency.

Don't let your gas tank get low. Carry a can of gas if you can strap it to the outside of your vehicle, where it can vent.

Some people suggest storing canned dog food as your emergency food supply so you won't be tempted snack on it casually.

Carry with you at all times

❑ Jackknife with can opener and other tools

❑ Pocket flashlight

❑ Pencil

❑ Change for pay phone

❑ Note pad

Maintenance

Making preparations should encourage peace of mind, but don't forget to follow up. Routinely check food provisions to ensure adequate shelf life. Hold refresher "duck, cover, and hang-on" drills. If that out-of-town phone contact moves, remember to issue a new number. And keep abreast of government planning, because location of evacuation routes, homeless shelters, and distribution centers may change.

For more helpful lists, charts and planning aids to help make your home and office safer in the event of an earthquake, turn to DISASTER PLANNING AIDS, page 141.

Beware of potential danger, such as being too close to older brick buildings, when the shaking starts.

Chapter 4
RIDING OUT AN EARTHQUAKE

A t Chenega, a seventy-foot wave followed the earthquake, some said within ten minutes; some said even sooner. It wiped out the village and killed twenty-three persons. Spray from seismic sea action hit the schoolhouse (standing on a knoll ninety feet high)... Most of the victims were caught in their homes or along the beach... The first wave caught a father carrying two small daughters while a nine-year-old ran beside him. The wave took the nine-year-old and tore one of the other children from his grasp... A woman survived a savage buffeting by the waves that stripped her of a long winter coat and all other apparel except one anklet.
— from "The Great Alaska Earthquake of 1964," a report by the National Academy of Sciences

Casualties from earthquakes are seldom from the gaping fissures in the earth that are favored in Class-B movies. Deaths usually result from other, equally deadly, conditions: tsunamis (as at Chenega), landslides, snow avalanches, failed dams, collapsed bridges and buildings, and fires. In this chapter, we'll first take a look at two of North America's most dramatic earthquakes—the 1964 Alaska quake and California's 1989 Loma Prieta temblor—and then go over a variety of things you can do to help you and your family ride out an earthquake.

The classic Alaska shocker

Alaska's Good Friday quake of 1964, the most violent ever recorded in North America, registered 8.5 on the Richter scale. Most of the death toll (131) was due not to the earthquake itself but to sea waves caused by the quake.

National Oceanic and Atmospheric Administration

A quake-caused landslide destroyed 75 homes in Anchorage when underlying ground slid toward Cook Inlet in 1964.

March 27 dawned snowy in Anchorage but clear and relatively mild in southwestern Alaska. Many on Kodiak Island remember it as singularly lovely and calm day. But no one reported a disappearance of wildlife, as sometimes happens just before quake activity. No premonitions were recorded, even in Valdez, northernmost ice-free port in America, only fifty miles east of the earthquake's epicenter. It was business as usual in the historic little gold rush town of 1,200 until just after 5:36 p.m.

The big event for Valdez that afternoon had been the arrival of the coastal supply ship Chena with a delivery of Easter lilies. A group of twenty-eight adults and children were gathered at the dock to watch it unload when the quake struck with an unbearable rumbling and the eerie, violent ringing of the bell atop the Church of St. Francis Xavier.

The shore began to roll, forming wavelike crests and troughs somewhat closer together than those found at sea. The pressure sent up twenty-foot spouts of water, sewage, and sand, and fissures opened and closed as the ground waves passed.

Jim Bedingfield, owner of the Valdez Hotel, abandoned his pickup truck only to fall into a deep fissure that opened under his feet, but the moving force pitched him right back to the surface. Downtown, buildings tottered on the land swells, but it was the waterfront that was hardest hit.

According to the National Academy of Sciences report:

The harbor became a maelstrom and the big dock began to break up; mounds of water hit the Chena. When Captain M. D. Stewart reached the bridge [of the Chena], the ship was lying over to port 50 to 70 degrees. The noise was tremendous, and witnesses saw incoming waves raise the freighter thirty feet, higher than the dock's warehouses. Captain Stewart looked down to see people running on the dock, but as they ran the dock disappeared. The warehouses, the packing plant, the cannery, the bar, the people plunged into the boiling seas.

Ten minutes after the first wave, a second followed. A total of thirty-one people died, including twenty-eight who disappeared from the Valdez dock. Captain Stewart and his crew managed to save themselves and their ship, however, and their radio operator became the town's sole communications link to the outside world through the turbulent night that followed.

Two were killed in 1964 when a precast concrete wall fell from the J.C. Penney Building onto nearby parked cars.

This Alaska roadway was left with a deep crack running down the centerline as ground shifted in the Alaska quake.

Farther west, the population center of Anchorage (50,000) was spared lethal tsunamis and runaway conflagrations. The death toll there was only nine: one man found under downtown sidewalk rubble, two children and several adults lost in earth slides, and one man dead from a brain injury. Yet the town was unusually hard hit, according to the National Academy of Sciences report:

> As the shaking continued, many of the Anchorage bluffs began to break off and move forward in "slides." Along downtown Fourth Avenue, the ground broke away in an irregular line in front of a row of business establishments; the buildings sank straight down as the slide followed the slope behind them. A service attendant farther along the street fought to keep his footing while he watched two and a half blocks of shops, bars and stores slowly settle until their entrances were below street level. The marquee of the Denali Theater came to rest on the street surface.
>
> However, none of the bluff failures was so devastating as the earth movement in the Turnagain Heights subdivision where some of the city's most prominent families had been relaxing before dinner. Within seconds, the shaking swelled into a savage grinding roll. The clay formation under the bluffs broke loose, and the ground gave way in a concave pattern along 4,300 feet of the choicest property in the subdivision. Houses, streets, trees, cars in driveways,

everything on the bluff side of the breakaway begin moving out and down toward the tide flats of Knik Arm. Members of families were separated in the moving mass; crevices opened in the already disrupted earth, and clay ridges ten to twenty feet high emerged, completing the destruction of some seventy-five homes, parts of which rode in the slide down more that thirty-five feet below the old bluff level.

In downtown Anchorage, thirty blocks of dwellings were destroyed or severely damaged; the split-level Hillside Apartments were warped and sheared, wrecked beyond repair; the newly opened five-story J.C. Penney store twisted, sagged, and dropped a curtain wall of precast panels on the street.

U.S. Geological Survey

A span of this former railroad bridge dropped into the Copper River during in the Alaska quake.

Yet just two blocks south, 700 children watching a Disney film in the Fourth Avenue Theater rode it out unflinchingly after manager Joe Marboe "talked sharply" to them.

"I believe [the children] were more scared of me than they were of the earthquake, because we had no panic. We had no one leaving their seats during the whole time," he recalled.

The deeply footed Westward Hotel just a block away also survived pretty much intact, but a witness reported the 14-story building swayed wildly and metal doorways (traditionally considered a safe earthquake haven) bowed in and out with the impact. Fearless glacier pilot Bob Reeve, who was enjoying a birthday party on the top floor, decided evacuation was appropriate.

California differences

While the 1964 Alaska disaster has yet to be matched for physical impact anywhere in North America, subsequent California quakes of lesser magnitude have far surpassed it in damages. Estimated losses caused by the Good Friday event in Alaska (8.5 magnitude) were about $321 million, while a 6.4 quake in the San Fernando Valley (1971) cost about $511 million and the 7.1 Loma Prieta disaster (1989) cost some $6 billion.

The reason, of course, is population density. In 1964, Alaska was America's most sparsely settled state, with only 50,000 residents in its most populous city. Though a few tall buildings toppled in

Anchorage, Alaskans did not have to dodge flying glass from super high rises because none had yet been built. They did not have to worry about the dangerous bric-a-brac of Victorian architecture—cupolas, clock towers, gargoyles, and ornamental drain pipes—because Anchorage was built after that era. Also missing was the California hazard of collapsing elevated freeways, for Alaska had none at that time.

An overpass collapsed and nearby highways sustained heavy damage in the San Fernando earthquake of February 1971.

A dress rehearsal for the Big One

Many quake experts like to point to California's 1989 Loma Prieta quake as a "dress rehearsal for the Big One" which has long been predicted for America's most fault-riddled state.

And despite huge financial losses, Loma Prieta remains a classic example of how to face a major seismic event with ingenuity and courage.

Luckily, the epicenter was in the sparsely populated Santa Cruz Mountains. But a scant fifty miles north in the densely settled San Francisco Bay area, a 1.4-mile stretch of the double-decked Nimitz Freeway collapsed into a smoking ruin, crushing cars and sandwiching dozens of commuters in debris.

As the remains of the structure teetered from aftershocks, Dr. James Betts and several others inched through a narrow gap to the aid of six-year-old Julio Berumen, who lay trapped by the corpse of his mother's friend. The rescue, which took several hours, required dismembering the corpse with a chain saw and amputating the child's right leg at the knee, yet those involved say they never thought twice about seeing it through.

Others who stayed with the macabre dismantling were rewarded with the discovery of Buck Helm, age fifty-seven, a dock supervisor buried alive for ninety hours in the wreckage of his compact car. After five hours of frantic digging by rescuers, Helm was carried to safety, gamely waving to well-wishers, only to die about three weeks later of heart failure in the midst of negotiating movie contracts.

Baseball fans around the country witnessed the beginning of the Loma Prieta quake because the World Series between the San Francisco Giants and the Oakland Athletics was being broadcast from San Francisco's Candlestick Park at the time. The stadium, which had recently been reinforced, sustained little damage. But nearby, apartment buildings crumbled and fires roared. Death toll in the Bay Area eventually reached sixty-seven and some 14,000 people were displaced—but a worst-possible scenario had projected 4,000 deaths, so city residents counted themselves lucky.

How to ride out a quake

Writer Joan Didion has often wondered at the seemingly blase attitude of longtime Californians to the earthquake dangers of their Golden State. In her book *After Harry,* she attributes it to "protective detachment, a useful adjustment commonly made in circumstances so unthinkable that psychic survival precludes preparation." But times are changing. Witness these developments:

Architects have developed building stabilizers that can counterbalance quake movements and help keep high rises on an even keel.

Federal regulations now require even high school chemistry labs to systematically store supplies so they will not form lethal mixes and chemical clouds during earthquakes.

A California firm, noting an extraordinary number of quake-sustained injuries among high-rise-based workers who couldn't eject themselves from fast-rolling office chairs before smashing their knees into walls, has developed a chair caster that self-arrests.

And most Californians have now had some form of earthquake training at school, on the job, or via television that makes "duck and cover" a reflexive action.

Surviving

For the best chance of riding out an earthquake, whether you are indoors or outside, there are some important tips you can follow.

Indoors

When the shaking starts, duck and cover. At home or in your office, get under a solid, sheltering piece of furniture like a table or a desk. If it moves, hang on and go with it.

Avoid rooms with large ceiling areas, hanging fixtures, mirrors, and glass. Also avoid kitchen areas where large appliances may topple and pin you down. Doorways can be a haven if they're attached to the main structure, but even some metal doorways can buckle under stress. If they have doors that can swing in on you, stay clear.

If caught in a public building, steer clear of panicking crowds as well as structural hazards. In a theater, you'll probably do better to crouch in the protection of your seat than to fight the mob heading for the exit. The same is usually true at a ball park or stadium; your chances are better in the open playing area than jammed at an exit.

Luckily, this six-story apartment building in Anchorage, made of concrete slabs, was unoccupied in March 1964.

Upper-floor walkways can be particularly dangerous at shopping malls, where flying window glass will also be a hazard. Head to the back of a shop; go to clothing or yard goods displays rather than appliance sections. One of the all-time worst places to be is in a big discount house where heavy items are stored high overhead. If you can't get clear, find something large and solid to crouch under, or at least beside, protecting your head.

Outdoors

Stay clear of toppling building walls, street lights, trees, utility wires and poles, and slide areas. If driving, it's usually best to stay with your car, which should give good protection unless you're beneath a bridge or freeway overpass. Drive to get clear, if possible, then pull off the road, setting your emergency brake.

Watch your head

Once you are safely positioned, weathering a rocker is a matter of the right psychology. Some therapists recommend taking charge, giving yourself the feeling of being in control by demanding—aloud and with authority—that the quaking stop. This approach will certainly keep you occupied, but it's well to remember that it's also likely to antagonize fellow victims not of like mind who, for example, may feel that praying is a more appropriate response.

Another school of thought is to occupy yourself with something you truly enjoy. Tenor Enrico Caruso, a reluctant witness to the great San Francisco quake of 1906, sang his way through the worst of it. A couple who survived the San Fernando Valley quake of 1971 told the *Los Angeles Times* that they had engaged in some mind-diverting maneuvers on a water bed and agreed that seismic action heightened the experience.

Naturalist John Muir, who went through a California temblor in 1872, discovered he enjoyed the experience in itself:

"At half-past two o'clock of a moonlight morning in March, I was awakened by a tremendous earthquake, and though I had never before enjoyed a storm of this sort, the strange, thrilling motion could not be mistaken... Both glad and frightened, [I shouted]: 'A noble earthquake! A noble earthquake!' feeling sure that I was going to learn something."

Mountain climber Bill Mattison didn't have time for much enjoyment as he weathered a 6.0 quake near its epicenter on Mount McKinley in May of 1991. "It really jostled us around," the Vail, Colorado, resident told the McClatchy News Service. "I just grabbed my [sleeping] pad and tried to hang on. When I unzippered the tent, the whole entire north face of [Mount] Hunter was coming down. It looked like a rapid in a river."

Perhaps the most positive approach—in addition to camping beyond the reach of avalanches—is to keep in mind that your chances of surviving a quake today are much improved over years past, thanks to increased knowledge of earthquake behavior and to the kind of preparedness advice outlined in this book.

Chapter 5

AFTER THE QUAKE

The most important decisions for survival in seismic disasters are often made just after the quaking stops. The first challenge is to assess your location and decide whether it's safe to remain there. To do this effectively, you need to make some unpleasant assumptions:

1. There will be aftershocks.
2. If you live near the sea, tsunamis (large sea waves) will follow.
3. You will have no access to medical or police assistance for as long as ninety-six hours.

Perhaps none of these assumptions will turn out to be accurate. But to be safely prepared, you must act as if they are.

Aftershocks

While many seismic events are simply a single jolt, those over magnitude 7 on the

Richter scale typically spawn about six aftershocks of magnitude 5 or more. These may occur any time from less than a minute after the initial shock to more than a year later.

A 7.4 quake that hit the Mojave, California, area, on July 28, 1992, set off a second quake registering 6.5 three hours later. The two quakes, occurring on different faults, were followed by more than a dozen aftershocks registering magnitude 5 or over.

Aftershocks may set off fire and burglar alarms and sprinkler systems. They can also be the straw that breaks the camel's back in quake-weakened buildings.

During California's 1989 Loma Prieta quake, scientists at the U.S. Geological Survey transmitted radio warning signals at the instant that large aftershocks struck. The signals provided warning to rescue crews tens of miles away, several seconds in advance of the onset of strong shaking (radio waves travel much faster than earthquake waves). Workers were then able to avoid damaged buildings during the worst of the aftershocks.

Tsunamis

Following a sharp Aleutian quake in 1946, a 100-foot wave struck Scotch Cap on Unimak Island, Alaska, wiping out a lighthouse with its crew of five men. Four hours later, an undulation about four feet high gently rocked two freighters moored one mile off Hilo Bay, Hawaii, 2,000 miles away. Then, as crews watched, horrified, a mighty wall of water suddenly rose from the ocean's surface to crash fifty-five feet up on the Hilo shore. Death toll was 173, the worst natural disaster in the history of the island.

Tsunamis overturned this ship and left much of the Seward waterfront in ruin after the Alaska quake.

The disaster prompted establishment of the Pacific Tsunami Warning Network in use today. Be aware, however, that this system may be too slow to help those near the epicenter of a seismic disturbance. And it cannot be relied on to predict locally generated waves, as was demonstrated in the Alaska quake of 1964.

In the Aleut village of Kaguyak (population thirty-six) on Kodiak Island, radio operator Joe

Melovedoff spread the warning as soon as he saw water coming in following the 1964 quake. All the villagers made it to safety on a fifty-foot hill, and the first wave did little damage. A second and third tsunami struck. Villagers assumed the third wave was the last—but a fourth, much the largest, struck from both sides of the bay.

Walling in the village, the fourth tsunami formed several whirlpools, lifting the houses and church off their foundations. It caught five villagers and a visiting geologist and his wife. The geologist's wife and two villagers escaped; the others were killed.

When it comes to tsunamis, public curiosity often overcomes prudence. The 1964 Alaska quake claimed eleven tsunami victims in Crescent City, California. Most of these were people who had returned to the beach after the first wave. A crowd of 10,000 people waited on San Francisco beaches, which were fortunately bypassed by high water. British Columbia's first official tsunami warning, following an Aleutian quake in 1986, attracted hundreds of spectators to the beaches. It turned out to be a false alarm.

It is difficult to imagine the impact of a tsunami. The speed of seismic sea waves is determined by the depth of water through which they pass. They can travel more than 600 miles per hour in deep water but slow to 30 or 40 miles an hour when approaching shore. In deep seas, wave height from trough to crest may be only a few feet, not particularly noticeable from the air or aboard a ship. However, as tsunamis are slowed by shallow water, they pile up, increasing in height, sometimes reaching 100 feet or more. Approaching crests are sometimes preceded by a roaring sound. They may also be heralded by a noticeable rise in coastal water or drop in water level, even exposure of the ocean floor.

The waves in a tsunami often come in a sequence, spaced ten minutes to half an hour apart. As Alaskans discovered in 1964, there may be four or more of widely varied intensity. A small tsunami at one beach can be a giant a few miles away. Expect the unexpected. If possible, stay away from the beach until you hear an all-clear from the warning center. And keep in mind the first rule of tsunamis: when you can see the wave, you are probably too close to escape it.

Lack of aid

Because big quakes often destroy transportation and communication networks, citizens may find themselves on their own when it comes to evacuation, fire fighting, and rescue operations.

During the Loma Prieta earthquake, selfless volunteers performed with remarkable courage and ingenuity, saving dozens of lives. But use extreme caution in undertaking such free-lance ventures. Collapsing buildings and shifting rubble took the lives of about 150 rescue workers during the Mexico City quake of 1985. Earthquakes that produce massive aftershocks can be particularly dangerous.

During the wait for outside help, survivors sometimes organize their families and neighborhoods to handle assignments like sanitary services and assessment of damages. Neighbors standing together can also keep looters at bay. Taking charge helps to restore the feeling of being in control that a big earthquake takes from its victims.

Since electronic banking units usually crash and money may be useless in disaster areas, a barter system sometimes develops with surprising priorities. Earthquake lore is rife with stories of people swapping diamond rings for jugs of water, cans of Spam, or a six-pack of beer. Sometimes a cache of trading provisions can be invaluable in survival, especially if an orderly exchange plan can be worked out.

Surviving after the quake

In the aftermath of an earthquake, you can increase your chances of injury-free survival by taking positive steps to help yourself and by sidestepping some avoidable dangers.

Wearing stout shoes to guard against broken glass, check your earthquake refuge to see if it remains structurally sound. Pay particular attention to the foundation. Proceed with caution. Use a flashlight for illumination; do not light any matches until you are certain there are no gas leaks or combustibles nearby.

Be aware that stairways—especially in high rises, flimsily constructed apartment buildings, and shopping malls—can be death traps. Elevators, although usually secure from falling, may stick. Underground parking lots are prone to collapse.

If utilities are still functioning, don't turn them off unless absolutely necessary. Too many gas-line shutdowns at one time could rupture the main. But if you detect leaks or suspect (smell) hot wiring within the walls, waste no time in getting to the shut-off switches.

Do not flush toilets until you are certain sewer lines are intact. Besides, you may need the water in the toilet tank (and in your water heater) for drinking and cooking. (See Chapter 3, "Prepare Now!," for information on how to purify such sources of water.) Don't use swimming-pool water for drinking because the salts and acids used in pool maintenance can cause kidney damage. If your water lines still function, fill your bathtub as an extra water supply in case future quakes cut you off.

Ration all supplies until you're certain more will be available. Start using the food in your refrigerator first, because it will be the first to go bad if the electricity fails. Keep your freezer closed as long as possible to conserve the cold.

Remember that too many phones in use at once can bring the whole system down, which is why you should refrain from making calls unless absolutely vital. Instead, turn on your radio to get an idea of how widespread the damage is (keeping in mind that announcers may be unnerved, lacking vital information, or reacting to false rumors). In areas where disaster planners have prepared well, radio information can be invaluable.

Traveling and other dangers

If you decide to search for loved ones, leave a note showing your proposed route, because chances are they will come searching for you. Lock up valuables and secure your house as well as possible before you go. Leave food and water for pets if you can't find anyone to look after them because you may not be able to return as soon as planned.

People who live on low coasts, or in the shadow of dams, nuclear power plants, or volcanos, usually have planned evacuation routes worked out by civil authorities. In California, the Sixth Army has established corridors for safe travel. But evacuation can be a tough decision following an earthquake because, generally, you will be better off to stay where you are—even in case of medical emergency.

With its large freeway system, California faces greater potential for quake damage than other western states.

People who managed to navigate quake-twisted freeways to get to hospitals during the San Fernando Valley quake of 1971 discovered that that some of the structures had collapsed and that 80 percent of the medical staff had deserted. Survivors would have done better to stay at home.

Driving may be impossible. The short but vital Hollywood Freeway has sixty-seven overpasses, and failure of any one would close the route. Many streets in Los Angeles run over potentially explosive subterranean gas lines. Equally risky are routes passing oil storage tanks which, when weakened, are potential fireballs. Travel cautiously in landslide areas, on roads that are down-slope from water tanks and dams, and on routes with overhead electrical wires. Drivers should also keep an eye out for unusual cracks and dips in the road, packs of frightened animals, and chemical clouds that might form over school and commercial laboratories and toxic waste disposal areas.

You may have a better chance of getting through if you travel by bicycle, motorcycle, all-terrain vehicle, four-wheeler, or snow machine. But the price is increased personal exposure and the risk of having your vehicle stolen.

Looters can be expected, especially in commercial areas and wealthy neighborhoods. Civil authorities or neighborhood watch groups formed to quell the looting may be just as dangerous. (It's suspected that more people were shot and killed as looters in San Francisco in 1906 than died as a direct result of the quake.)

Getting back to normal

When relief workers arrive, they may be expected to furnish medical help, heavy equipment for rescue work, food, clothing, even toothbrushes and sanitary supplies. Some agencies will help with making welfare inquiries and unsnarling the red tape required for disaster loans. Mobile telephones are often set up for emergency calls. Cash may be flown in to central locations for cashing checks and banking. Free evacuation may also be offered (but check to see if you have to pay for your return).

Getting back to normal life involves more than just providing for the basics of food, clothing, and housing. It also means dealing with the psychological damage from a quake. During the San Fernando Valley quake on February 9, 1971, the home of Dr. Stephen Howard "shook like a freight train" but survived. Howard evacuated his family, but they returned home that evening. Howard's wife, Jackie, who had survived World War II bombings in Austria as a child, found herself unnerved by the aftershocks. Both noticed their nine-year-old daughter was not responding well.

Howard, who was director of the San Fernando Valley Child Guidance Clinic, compared notes with a colleague, psychologist Norma Gordon, whose ten-year-old son appeared equally upset. They decided that many other children might also need help. A total of 350 families quickly responded to their offer of counseling.

Howard and Gordon found that despite some parents' fears that talking about the quake would increase their children's anxiety, exactly the opposite was true. They prepared a pamphlet on the subject, "Coping with Children's Reactions to Earthquakes and Other Disasters," published later by the American Red Cross and the Federal Emergency Management Agency (see section on Sources, at the back of this book).

"Once parents realized that talking helps, there was no problem," Howard recalls. "We encouraged parents to admit they were afraid, too. They had been telling the children, 'We're not afraid and you are.' The children suffered anxiety, believing something was wrong with them... that they weren't brave enough."

Adults, too, should realize they may be close to the edge and should talk things out. Whether or not you continue to live in a quake area will be an important decision, requiring serious thought. Counseling may help. Some areas are regarded as earthquake free (see Chapter 1, titled "At Risk"). But before you seriously consider relocating, do some careful homework, because these places can be prone to other disasters like tornados, hurricanes, and floods.

Should you elect to remain living where you are, you'll do so with the confidence of an experienced earthquake survivor, and the value of that asset should not be overlooked.

Chapter 6

SAFER BUILDING

Advances in designing to survive earthquakes and a growing emphasis on strengthening existing houses and other structures is slowly bringing us safer buildings. In this chapter, we'll hear about the earthquake engineering that is creating big buildings that are more earthquake-resistant. Then we'll look at the steps *you* can take to make your own home safer.

Earthquake engineering

In 1979, Atlantic Richfield Company (ARCO) embarked on building a twenty-two-story tower in Anchorage, Alaska, the city that was hit with the worst quake in North American history fifteen years earlier. After taking soil samples from all over town, company officials decided to build the tower atop a two-story structure erected in 1965, the year after the big quake.

The result is an impressive tower built to

sway as much as six feet to help mitigate quake impact. The steel frame is mounted on springs. Boiler, air conditioning unit, pipes, and furnaces are bolted down. The biggest shock the tower has weathered to date is a magnitude 6.4 shake, comparable to the devastating 1971 San Fernando Valley quake. The only damage was to dry wall on the tower's second floor, ARCO reports.

Bemused survivors who have weathered major seismic disturbances on its upper floors describe the experience as "memorable" and "moving," and most voice confidence in the tower's unique design.

"A friend in a neighboring building tells me she'll know they are in trouble when they see our building go floating by," says ARCO public relations officer Sue Andrews.

While it's a misnomer to call modern high rises, like ARCO's, "earthquake proof," construction techniques have come a long way in a short time. Until 1957, engineers for the City of Los Angeles thought it implausible that high rises could be built to survive major quakes. City codes limited most structures to 150 feet, or about thirteen stories. But today one of the tallest buildings in Los Angeles is the sixty-story, 856-foot tower belonging to First Interstate Bank, a leader in earthquake preparedness.

The death toll from California's 1989 Loma Prieta temblor of 7.1 magnitude was sixty-seven in the Bay Area, with the worst damage confined

Houses and other structures not bolted to their foundations are vulnerable to destructive shifting such as this.

to some apartment buildings, a fifty-foot section of the San Francisco-Oakland Bay Bridge, and a long section of elevated freeway. By comparison, a rocker of somewhat smaller size (6.9) in December 1988 killed 25,000 people and leveled whole communities in Armenia, where modern building techniques are virtually nonexistent.

"The obvious lesson is that stronger structures save lives," *Newsweek* magazine editorialized in a feature on architecture titled "Designed for Survival."

A key to architectural earthquake survival is better understanding of soil conditions and of the intricacies of harmonics. When presumed "earthquake proof" buildings collapsed during the Mexico City quake in 1985, much of the blame was placed on the fact they were built on the soft, unstable ground of a dry lake bed or over sections of the city's vast subway system. A major problem in the Anchorage quake of 1964 was liquefaction, in which moist clay shaken by seismic waves flowed like a liquid. Liquefaction also was a culprit in the Loma Prieta quake, where unstable landfill became fluid.

An understanding of harmonics is also helping answer questions about building integrity. "A building is like a tuning fork," architect Scott MacGillivray explained in a *Los Angeles Times* interview. "When you put two tuning forks next to each other and strike one, the other will start ringing. The same happens in an earthquake. If the building is harmonically in tune with the earthquake, the building will shake more." But according to MacGillivray, buildings can be programmed ("tuned like a piano") so that their resonant frequency is not a natural harmonic of the predicted earthquake wave and they will remain relatively stable.

While early attempts at "earthquake proofing" were predicated on building solidly for strength, engineers have discovered that chances of survival are better for tall structures if they sway and absorb shock. The five-dollar word for this phenomenon is "ductile," referring to a combination of flexibility and strength—the ability to bend but not break—found in the steel frames of skyscrapers and the steel reinforcing bars (re-bar) placed in modern masonry.

Base isolation, a technique tested overseas and involving the building of shock absorbers into foundations, also appears to be gaining popularity in North America. For small, box-like buildings, its use can actually cut costs by 5 percent (in addition to the possibility of cutting insurance costs), according to Douglas Way, vice president of Base Isolation Consultants. Unfortunately, in the majority of cases, base isolation adds as much as 15 percent to construction costs. But experts predict it will soon be commonplace because of its effectiveness.

Even more revolutionary is the design of "smart" buildings equipped with weights, flexible cables, and computerized sensing devices geared to maintain a "sense of balance" during seismic events.

Following the Loma Prieta quake, the Federal Emergency Management Agency noted that public elementary schools, high schools, and community colleges generally held up well. However, an improvement program for bridges and freeways, on which California had spent about $54 million, did not rate as well, and collapse of part of the double-decked Nimitz Freeway sent designers back to the drawing board.

In large part, earthquake safety is a matter of economics. When upgrading for a once-in-a-lifetime event can cost 20 to 40 percent of a structure's replacement value, planners must carefully weigh their decision. Often they decide to go with the risk.

But the price of mitigating earthquake risks may lessen. Elaborate quake-proofing—like the computer- controlled, motion-sensitive weights and pulleys of "smart" buildings—seems prohibitively expensive today. Yet this developing technology may ultimately allow engineers to design more economical buildings using lighter, more flexible materials, and costs could even out.

Making your home safer

High-tech experimentation is not necessary for the average home owner. According to a U.S. Geological Survey publication produced after the Loma Prieta quake, "Most people in California are safe at home if they live in a one- or two-story wood-frame building. These buildings are not likely to collapse during earthquakes. The most common damage is light cracking of interior walls, cracking of brick chimneys, and cracking and possible collapse of brick veneer on exterior walls." The same applies to log cabins, which showed a good survival rate during the Alaska quake of 1964.

Of course, the odds can change if your house is located on loosely packed ground or wet soil subject to liquefaction. You may also have cause for concern if your house was built before 1940, if it is constructed of unreinforced masonry, or if

Look For the Danger Signs

Be aware of structural hazards that can spell disaster when an earthquake strikes

Large tree close to house

TV antenna attached to weak or old chimney

Heavy tile roof

Unstable TV satellite dish

Chimney too close to exit

Heavy plastering on ceiling

Poorly reinforced masonry walls or fences

Neighbor's house too close

Lots of glass

Large overhang

Hillside location; housing on stilt-type support

Split-level construction

it is built with too few bearing walls to resist the horizontal forces of a quake. And don't make the mistake of thinking that just because an old house survived the last quake, it is safe. It may well have suffered internal damage that will make it doubly dangerous.

The good news is that there are ways to drastically reduce quake risks by reinforcing vulnerable dwellings. Statistics show that less than half of 1 percent of all reinforced units had to be vacated during the Loma Prieta quake, compared with 5.1 percent for buildings that had not been "quake-proofed."

The practice of using the first floor for a garage may leave inadequate lateral bracing on the ground level.

The first step in assessing the safety of your home or one you are considering purchasing is to find out what kind of ground it is built on. Most

California Realtors have maps of their areas showing filled land, known faults, and potential liquefaction or landslide areas. If no such maps exist, consult an engineering geologist, geotechnical engineer, or foundation engineer.

Next determine what your building is made of—which may not be as simple as it sounds. Sometimes what appears to be concrete, masonry, or brick is actually wood frame. And stucco—a combination of portland cement plaster applied over wire mesh—can be wood-based. If a knock sounds hollow on the inside walls of a stucco building, you're actually dealing with wood. A hollow sound coming from the exterior of brick, stone, or cement will also indicate wood framing behind a masonry veneer.

In general, the newer a building, the safer it is likely to be. But don't go on looks alone. Old buildings can be renovated to look modern. Age should be checked through the tax assessor's office, a historic Sanborn map (published early-on for most American cities), building permit file, or even utility records.

Among older wood-frame houses, two structural problems are common. Of the more than 23,000 homes damaged in the Loma Prieta quake, most had no anchor bolts to attach their walls to the foundation or lacked adequate bracing for their cripple walls (short walls between the foundation and the first floor). Both problems can be remedied (see accompanying illustrations).

The force of a quake is seen in these modular units (above) whose support jack (bottom) has buckled.

California's Unreinforced Masonry Building law required that all local governments inventory masonry structures and develop a mitigation plan by 1990. Statistics show that those built since the mid-'70s usually perform well in quakes, while earlier models require reinforcement. Mortar quality should also be taken into consideration when dealing with brick. Some early California mixes were so weak that they earned the name "buttermilk mortar."

Adobe and stone buildings, particularly those without reinforcement, are designed to carry vertical loads only. These buildings have almost no flexibility or strength in resisting the lateral forces of a quake and are the least easily rehabilitated.

In trying to assess the earthquake risk to a house, make a note of these potential danger signs:

- Split levels and complex geometry
- Large amounts of glass, windows, and doors, particularly at building corners
- Inadequate bearing walls in garages that support living quarters above
- Stilts supporting structures, such as on a hillside site
- Large overhangs
- Weak chimneys
- Tile roofs
- Plaster ceilings (which can weigh as much as eight pounds per square foot, with a room fifteen feet square supporting as much as a ton of ceiling)
- Poorly braced television antennas (especially attached to chimneys)
- Poorly braced satellite dishes

❏ Poorly engineered masonry walls (free-standing or retaining; structural or for the garden)

❏ Trees that might topple on your dwelling

❏ Buildings or pools poised uphill of your dwelling

❏ Neighboring houses that are unusually close

❏ Unusual features and configurations in a house are not necessarily hazardous, engineers note, but they can be vulnerable to damage if poorly designed or constructed.

Trailer homes generally weather quakes well if they are on wheels or if they have a good structural support bracing system. Mobile home dealers and trailer home associations can often supply information about seismic bracing systems.

Investing in an upgrade

Well-planned earthquake upgrading will add value to your home, protect your investment, and help keep your family safe. Of course, it also can be inconvenient, noisy, messy, and expensive. You may help lower costs by planning quake upgrades in conjunction with other home improvements like remodeling, repairing termite damage, adding insulation, or installing solar panels. You can also lower costs by doing the work yourself. The Federal Emergency Management Agency (FEMA), the Red Cross, and numerous private companies offer free detailed plans to guide you.

If your home needs extensive reinforcement, it may be wise to hire an engineer or architect to help with planning. The fee may represent between 10 and 15 percent of the total reinforcement cost.

Bolting Down the House

For homes not already securely fastened to their foundations

Drill holes through existing sill into concrete foundation for expansion bolts, using carbide drill bits. Utilize right-angle drill for tight access places. Use $\frac{1}{2}$" diameter bolts about $7\frac{1}{2}$" long, placed every 3 to 4 feet along the foundation.

NUT

WASHER

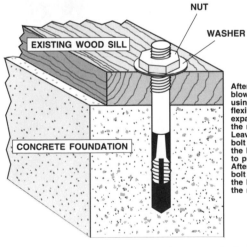

EXISTING WOOD SILL

CONCRETE FOUNDATION

After dust has been blown out of holes, using a piece of flexible tubing, insert expansion bolt with the nut attached. Leave nut at top of bolt when tapping the bolt in place to protect threads. After tapping the bolt in place, tighten the bolt by turning the nut.

Locating knowledgeable contractors may not be easy. The California Contractors State License Board has produced a free booklet, "What You Should Know Before You Hire a Contractor" (see the Sources section at back of this book). FEMA suggests soliciting bids from at least three firms, carefully investigating variations.

Be particularly aware of home improvement scams in recently quake-damaged areas where high-pressure, fly-by-night operators talk a good game but make only cosmetic repairs. One tip-off may be when the contractor contacts you (instead of you calling the contractor). Check with your Better Business Bureau or the statewide contractors association if in doubt.

Strengthening Cripple Walls

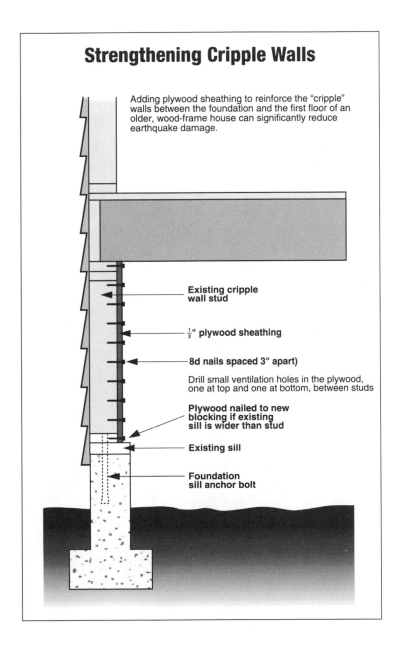

Adding plywood sheathing to reinforce the "cripple" walls between the foundation and the first floor of an older, wood-frame house can significantly reduce earthquake damage.

Existing cripple wall stud

$\frac{1}{2}$" plywood sheathing

8d nails spaced 3" apart)

Drill small ventilation holes in the plywood, one at top and one at bottom, between studs

Plywood nailed to new blocking if existing sill is wider than stud

Existing sill

Foundation sill anchor bolt

Chapter 7
EARTHQUAKE BUSINESS

Rubble had scarcely settled from the terrible San Francisco earthquake of 1906 when legendary financier A. P. Giannini set up business on the sidewalk and started loaning money for rebuilding. His foresight helped salvage a great American city, and his company, Bank of America, become one of the nation's most successful financial institutions.

Developing a disaster plan as part of your business plan makes sense, not just to protect your investment but to be ready for what might be a unique business opportunity.

"Why should we be victimized by something we know is coming?" reasons John P. McCann, former vice president of the Insurance Information Institute and now a California publisher. "Why should we fail at business when, in fact, we probably could have taken the right steps to be able to capitalize on the disaster. I don't want to make it sound like this is opportunism of the worst order when, in fact, it's just the opposite if we could motivate business to see its role in society here.

"Afterwards you're going to sell three million T-shirts with 'I survived the great quake' and all you've got to do is throw in the date, the people who are going to be supplying those shirts are going to have to be up and ready."

T-shirts may not be your specialty, but downtime affects the bottom line of even the most conventional business. The longer it takes your company to recover, the higher your costs. Experience with recent disasters indicates that if a business has not recovered within two weeks, chances are it won't recover at all.

A business survival plan

In quake-prone areas like Los Angeles, San Francisco, Anchorage, and Vancouver, major companies have developed extensive disaster plans. Many are also working with government agencies in order to make people in the public and private sectors aware that earthquake survival is possible. These activities represent enlightened self-interest. Business survival depends on the survival of both employees and customers.

Legal liability is also becoming a serious consideration, since business leaders can no longer claim they are unaware of earthquake dangers.

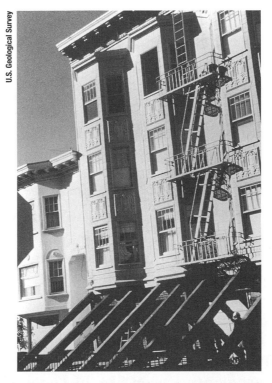

Increases in demand for certain goods and services may stimulate the economy after a destructive urban quake.

Previously, earthquake-related deaths and injuries could be dismissed by the corporate sector as acts of God, but courts today are less likely to be lenient on business owners in high-risk areas who make no attempt to prepare for an earthquake. Bluntly put, earthquake-injured employees and customers—and victims' heirs—may well sue your socks off if you choose to ignore the risk.

Happily, the price of earthquake preparedness is relatively low, and expenses should be tax-deductible. Preparation may also serve you well in other disasters and provide the bonuses of enhanced employee confidence and customer goodwill.

Assess the risks

The first step should be to assess risks and assets.

Risks may include structural weaknesses at your plant, chemical hazards, equipment and inventory that could slide or self-destruct, and losses if you're required to shut down.

Assets to look for include employees who know first aid or CPR, those who can operate a ham radio or do computer networking, and construction workers who may be good at operating rescue equipment. Your employees may also include volunteer fire fighters or cool-headed former police officers or military personnel who fully understand the value of disaster training.

Government agencies like the Federal Emergency Management Agency and private-sector agencies like the American Red Cross are ready and willing to work with your company free of charge. (See the Sources section at the back of this book.)

Secure your office or plant

Correct structural weaknesses in your plant if possible. If this is too expensive, plan to avoid danger areas and work out contingency plans. Secure heavy equipment and toxic supplies with straps, bolts, or even Velcro. Consider ways to protect yourself from fire hazards.

Stock up on emergency medical supplies, food, water, batteries, etc., and work out a system for dispensing them. (Review Chapter 3, "Prepare Now!")

Work out procedures for gathering, analyzing, and reporting earthquake damage. Have an accurate inventory list ready for the insurance appraiser.

Involve your employees

"A successful disaster plan is not merely a three-ring binder which lists the phone numbers and resources, but rather it is the result of the combined planning efforts of many departments and people," observed Pete Ashen, administrator of emergency service, Golden Gate Chapter of the American Red Cross.

Teamwork is even more vital in implementing your plan. If your employees have invested their time and energy in working out disaster strategies, they are likely to carry them out efficiently.

Put together a group of key employees who are interested in disaster planning. Then figure out the worst possible scenario—fires, chemical spills, collapsed stairs or floors, stranded elevators, power and water outages, incapacitated personnel—and attack the problems systematically.

Develop a working plan that will allow you to continue operation in the face of shutdowns due to failures of utilities, communications, transportation, or banking. Consider what you will need for twenty-four hours, seventy-two hours, a week, and a month. Remember to include some procedure for contacting employees about returning to work and for paying them during the disaster period.

See that at least two employees assume leadership roles on each floor. These "floor wardens" should be trained in what to expect in a severe quake (such as building sway), safety rules, damage assessment, and evacuation procedure. Assign vital functions such as protecting customers, dealing with the handicapped, and shutting off power and gas if necessary.

Distribute and post the emergency directions that you work out with your employees. It's also important to update your business survival plan on a routine basis, paying particular attention to staff changes and special training programs. And regular practice, including earthquake drills and quizzes, will be necessary to stay current.

Also be sure that your employees' families are well prepared. Workers who are not overly worried about survival on the home front will be more effective on the company's rescue and recovery team.

Develop an emergency management plan

Specify an alternate headquarters. Set up fail-safe communications. (Don't rely on cellular phones, because their systems depend on antennas and are quick to overload). Back up financial records, computer data bases, and software, and store them outside of your fault zone. Arrange for security and easy access to your facilities by key employees.

Develop a plan which includes emergency requirements for a quorum and a succession system for senior management. Include the conditions for granting emergency powers to the acting chief executive and the degree to which company officers may be held liable while acting under emergency laws. Make clear who can legally call a meeting.

Since major disasters have a ripple effect that touches all businesses, pinpoint suppliers that are crucial to your operation. Develop a plan to counter scarcity or a steep rise in prices. Also consider the consequences if your major customers are in the quake zone. Diversification may be the answer.

Review investment strategy

Carefully assess all the stocks and bonds your company holds in corporations, public utilities, or municipalities that may be affected by a disaster. Planners have predicted that a really big quake might shut down major stock exchanges. You should also realize that insurance companies may have to divest themselves of large blocks of securities, possibly depressing prices, in order to pay off claims from a major disaster.

U.S. Geological Survey

A tremblor damaged this home in California, where the insurance industry is required to offer earthquake coverage.

Insurance

In 1984, California (the only state so far to address earthquake insurance problems) passed Assembly Bill 2865 requiring all insurance companies to offer affordable residential earthquake coverage. Insurers complained that the law didn't permit them to charge enough to cover actual residential losses. They countered by increasing the deductible from 5 percent to 10 percent of the actual cost of the building. Still, the law offered a break for many home owners in quake-prone zones who previously could not obtain reasonably priced protection.

Unfortunately, in the wake of the home owners' insurance law, California businesses and municipalities were hit with rate hikes of up to 800 percent and doubling deductibles because lawmakers set no ground rules for the insurance industry. And the problem may be even more serious in quake-prone states where the state governments have done nothing. Many businesses go uninsured. Even in California where residential coverage is relatively affordable, only about 20 percent of home owners purchase quake insurance.

A major deterrent to insuring is the federal government's tradition of reimbursing earthquake victims for a portion of their losses. According to Robert E. Litan, an economist at the Brookings Institution in Washington, D.C., American taxpayers shelled out an estimated $17 each to cover damages of San Francisco's 1989 Loma Prieta quake, and there is no indication this trend will change.

"Uncle Sam could play hardball, giving legal notice that in the future the Treasury would help only those who help themselves," speculated *New York Times* economic columnist Peter Passell. "But the threat would probably not be credible: Americans are no more likely to deny relief to uninsured victims of an earthquake than they are

U.S. Geological Survey

In California, the odds are 1 in 5 that this property owner had earthquake insurance to help pay for repairs.

to deny medical care to injured motorcyclists who do not wear helmets."

But what if the Big One actually comes, bringing astronomical losses? A recent study by the Federal Emergency Management Agency (FEMA) estimates damages could easily reach $40 billion in California and even more in the unprepared Midwest.

Spokesmen for The Earthquake Project, an insurance industry consortium, predict that losses from a major urban quake would total at least $100 billion and would severely disrupt the U.S. economy. They suggest forging a partnership between the federal government and the insurance industry to establish a quake-dedicated reinsurance fund. Congress continues to debate the problem from time to time, and business people are advised to keep alert to any Congressional decisions.

Capitalizing on a quake

Getting back to routine business before competition can recover from a crippling earthquake is the major goal of most corporate managers, but a few entrepreneurs see impending earthquakes as a unique opportunity for profit.

San Francisco architect Don MacDonald made news in 1991 by designing a "quake bed" to protect sleepers during the Big One. It features a 600-pound steel canopy with room under the mattress for a fire extinguisher, water, chemical toilet, books, heater, and food. Selling price was set at between $2,500 and $3,000, "depending on demand."

MacDonald has also designed a 14-foot by 17-foot "starter home" estimated to cost between $15,000 and $20,000 (not including the lot), inspired by "earthquake cottages" built to house victims of San Francisco's 1906 disaster. The compact but charming one-room dwelling includes a loft.

Survival kits priced from $15 to $150 prove increasingly popular with each earthquake prediction. Following Loma Prieta, a company called Yuppie Gear International of Los Angeles developed a kit that included imported caviar, Ile de

France pâtè, Stolichnaya vodka, and a last will and testament, along with the traditional gas turn-off wrench.

Earthquake tours are usually a big draw following a major disaster. And Universal Studios markets a "heart-throbbing, palm-sweating, mind-boggling 8.3 on the Richter scale" re-creation as part of its studio tour.

Astonished residents along the New Madrid Fault (extending from Cairo, Illinois, through Missouri to Marked Tree, Arkansas) found themselves inundated by curious tourists and hordes of reporters the week of a predicted quake that never materialized in 1990. Sales soared for shotguns, generators, purified water, and freeze-dried foods. Someone made a bundle selling T-shirts proudly proclaiming, "It's Our Fault."

Author Curt Gentry turned his college thesis on California into a best-selling book called *The Last Days of the Late, Great State of California,* a fictional look back at the Big One that finally rocked the Golden State into oblivion. And a number of forward-looking lawyers have created handy ready-to-use formats for lawsuits against businesses that fail to provide adequate earthquake protection.

Your earthquake portfolio

The most creative entrepreneurial activity is credited to stockbrokers who have designed "earthquake portfolios" to be automatically bought or sold when a major quake strikes the area selected by their client.

Inspired by turn-of-the-century speculator Edwin LeFevre, John Hartt of Kemper Securities in Anchorage has created a provocative disaster investment strategy. LeFevre made big bucks selling Union Pacific Railroad stock "short" on a hunch, just hours before other investors realized the price would drop drastically due to the havoc wrought by the San Francisco quake of 1906.

Heavy equipment such as this crane on the San Francisco Bay Bridge are in demand following a destructive quake.

Selling short is a real gamble which involves selling—at the current price—stock that you have not yet paid for. The idea is to sell the stock at the current price but to pay for it a short time later when the price has dropped. If things go as planned, you make a profit—perhaps a killing. If not, you're a loser.

Hartt's plan features the selling short of quake-prone companies like California municipals, utilities and, maybe, Union Pacific. This is coupled with purchase of stock in corporations that specialize in materials needed for quake recovery (building materials, glass, computer replacements) and which traditionally rise in price following a disaster.

Hartt's analysis of market reactions following three of the nation's most costly quakes (California 1906, 1971, and 1989) is fascinating. In 1906, slow communications and the unprecedented scale of the destruction resulted in stock speculators taking a long time to realize they were in deep trouble with San Francisco holdings. The market actually continued to rise until three days after the disaster, at which time many stocks sank at breakneck speed and some disappeared without a trace.

Reaction was just the opposite during the 6.4-magnitude San Fernando Valley quake of 1971. Initial press reports were so frightening that stock in companies based in the area dropped immediately. But many recovered the next day following a better assessment of damages.

Investors were even less rattled by the Loma Prieta quake of 1989, even though it registered 7.1 on the Richter scale and was brought to the nation live via the television hookup broadcasting a World Series baseball game from San Francisco's Candlestick Park.

Aetna Insurance stock actually gained 2¾ points (apparently on the assumption that few were insured and that rates would soon go up because of the disaster). Standard Oil of California gained a point, despite a temporary pipeline shutdown and the fact that the extent of damages to its refinery in the area were unknown. Union Pacific dropped less than a point and rebounded the very next day.

Follow-up analysis showed that in general, insurance, timber, construction, engineering, and building product stocks rose. The success story of the week was Kasler Corporation, a San Bernardino-based company specializing in cement pouring and highway repairs, which soared from 2⅛ to 9⅞. Also favored were companies that offered emergency computer services.

Shares in some California utilities, savings and loans, and banks dropped. Among the losers were Pacific Gas and Electric, Pacific Telesis Group, Bankamerica, Wells Fargo, and California saving and loans (though none lost more than a point).

Even investors who don't care to speculate on earthquake portfolios may find Hartt's observations useful in analyzing companies based in fault zones. There is no guarantee that the market will behave the same next time around, however.

Brokers warn investors to look carefully at each company as a whole. Some companies that sustain quake losses may be sufficiently diversified to carry on relatively unscathed.

Chapter 8

HOW PREPARED IS YOUR GOVERNMENT?

Many governments spend a lot of money, time, and talk on earthquake preparedness. But how well are they doing? And what can we as citizens do to make sure our governments are dealing with the problem?

The role of government

Lawmakers and everyday citizens face a host of important questions in debating the role of government. Should the government force the expense of earthquake mitigation on those who have homes in danger areas? Should it require earthquake insurance? Should valuable school time be taken up with training for a quake that may never come?

From basic earthquake research and experiments in building construction, to preparing citizens for disaster and helping pay the bills for earthquake damage, government already is deeply involved. In the mid-1970s, the U.S. Geological Survey began monitoring several hundred experimental quake-resistant buildings with sensors paid for by the National Science Foundation. One of the most unusual monitoring sites is the Century Plaza towers in Los Angeles, forty-four stories high, with three triangular bases spread out in fairly equal proportions around a central core.

With development of base-isolation techniques (incorporating "shock absorbers" into foundations), structures like the medical center at the University of Southern California and the rehabilitated City and Country Building in Salt Lake City were added to the monitoring list. Then in 1992, the Science Foundation allotted $5 million for the study of electronically balanced "smart" buildings, still very much in the experimental stages.

The impact of earthquakes on buildings also is studied in laboratories which use the motorized concrete slabs known as "shake tables" to simulate the effects of quakes on different materials.

The U.S. Geological Survey has long studied earthquake faults and has added satellite photography to its tools for documenting and monitoring the faults. Both Canada and the United States are teaming up with mineral exploration crews from the private sector to locate and explore undersea fault areas.

National Oceanic and Atmospheric Administration

A view of the devastation along Fourth Avenue, Anchorage's main business district, in March 1964.

The Federal Emergency Management Agency, formerly known as Civil Defense, received much criticism for sluggish reaction to the 1989 Loma Prieta earthquake in California, which unfortunately came on the heels of a Puerto Rican hurricane that took much of the agency's energy.

"Not that the agency doesn't like to plan, assess and evaluate," wrote Michael L. Cook in an editorial in the *New York Times,* "watching its spokespersons on TV during the crises caused by hurricane Hugo and the California earthquake, one gets the feeling that's all they do." The agency has since regrouped and appears to be moving aggressively into areas newly deemed at risk.

Currently a main cause for concern is how to pay the costs of future huge earthquakes. The insurance industry's proposal for a tax-free, self-sustaining fund (consisting of insurance premiums) to assist states that work with disaster planning agencies is opposed by government spokesmen who argue that the U.S. economy can survive the impact of such a disaster.

The federal government spends only about $1 per citizen annually on earthquake preparedness, compared with $350 per person spent by the government of fault-riddled Japan, where residents are reminded of a 1923 earthquake in Tokyo and Yokohama that killed more than 100,000 people.

In the wake of recent quakes and new, dire predictions, Californians—already leaders in earthquake planning—are moving to become even better prepared. But disaster specialists fear even they may grow lax again as public funding gets tighter.

The problem in Alaska is a combination of tight funding and bravado, according to observers at the Geological Survey. Anchorage residents who survived Alaska's huge 1964 quake seem certain they can do it again without much planning—this despite the fact the population is now five times denser than in 1964. And despite the fact that citizens have grown more dependent on urban infrastructure, natural gas for heat, and electric utilities. Typical is Walter Hickel, a pioneering Anchorage developer who, as governor of Alaska in 1992, vetoed funding for school earthquake planning despite the fact it would be matched 50 percent by federal money because of tight state funding.

Improving disaster planning and updating local building codes are two issues citizen activists can influence.

Making waves on the local level

If earthquakes are a threat in your area, there are a number of things that you, as a concerned citizen, have a right to expect from your local government.

Most important is sound disaster planning. Next is a building code that recognizes relatively inexpensive—but still state-of-the-art—earthquake mitigation procedures. Neither of these basics comes easy where the public is not generally aware of quake dangers—certainly the case in areas newly named at risk like Oregon, Washington, Massachusetts, and New York.

Even citizens of San Francisco, site of the most devastating quake in U.S. history (1906) and still at high risk from the nation's most active fault, showed surprising lack of concern about quake dangers when surveyed in 1987, not long before the Loma Prieta quake.

"That's because earthquake preparedness does not have a high priority with housing, transportation, environment, crime and homelessness to worry about," observed Philip Day, then the city's director of emergency services. "An awful lot of Californians feel there's nothing they can do to prepare. That's not true."

Even if you are not a public motivator or a political mover and shaker, there are many ways you can help prepare your community.

Educate yourself and your family

Check out your local public and school libraries for earthquake information. Both the Federal Emergency Management Agency (FEMA) and the American Red Cross offer an impressive range of material (see section on Sources at back of this book). Some state disaster planning offices also generate area-specific material, such as the wonderful brochure on tsunamis (*"Tsunami! The Great Waves In Alaska)* published by the State of Alaska (see section on Sources). Industries like insurance, building, engineering, and banking produce good planning guides and are anxious to work with private citizens.

Check the soundness of your area's disaster planning

Call (or better yet visit) your local fire department, and ask to check out its disaster plan. If a plan is available, make note of public shelters near your home and workplace and identify escape routes. Note danger areas like underground natural-gas and gasoline lines, toxic waste dumps, laboratories, and atomic power plants. Drive the evacuation route to see if it makes sense. Share this information with friends, family, fellow workers, and neighbors.

If your fire department doesn't have a disaster plan, ask the people whom the agency answers to—your mayor, city council members, city manager—if a community disaster plan exists and, if so, why the fire department does not have it. Make the inquiry in writing and send a copy to your local paper. Follow up to make sure you get a clear, satisfactory answer.

Ask your school board the same questions and follow up accordingly. The California governor's Office of Emergency Services has produced a booklet titled *Earthquake Preparedness Policy: Considerations for School Governing Boards* that makes an excellent model (see section on Sources). Also highly recommended is the teaching guide for a course on earthquakes *(Earthquakes: A Teacher's Package for K-6),* produced by the National Science Teachers Association and FEMA (see Sources).

Work with the private sector, too

Check to see if your phone book has a section on earthquake preparedness. If not, suggest to your phone company that this would be a good public service. And be on the lookout for other private businesses and agencies that might cooperate with government on disaster planning projects.

If there is little knowledge of earthquake preparedness in your community, consider inviting an expert to speak on the subject to your Chamber of Commerce, Rotary Club, Lions Club, Scout group, interested neighborhood group, hospital auxiliary, or other group. Look to your fire department, city or state emergency planning office, Red Cross, or area FEMA representative.

Check out regulation of quake-sensitive operations

In addition to checking your government's disaster planning, you might consider investigating its regulation of quake-sensitive operations like atomic power plants and the storage and disposal of toxic waste. Dela Ephrom, writing for the *Los Angeles Times Magazine* in 1987 after speaking to quake experts, shows that this is not for the faint of heart:

"They mentioned that one nuclear power plant, Diablo Canyon, happens to be built directly on a fault, and another, San Onofre, is built right next to one. They also said that UCLA science laboratories were filled with dangerous, unsecured chemicals and that if, after the quake, we see coming toward us a blue (or did they say green?) cloud, the toxic waste of a chemical fire, we should move crosswind to the cloud."

One early protest in front of the San Onofre facility, complete with folksingers and a student dressed as the Grim Reaper, brought out curious power-plant workers, beers in hand, to see what the commotion was all about. Since then, federal investigation into drinking and drugs on sensitive jobs has stepped up, along with enforcement of new codes for proper chemical storage in laboratories. But continued public interest is necessary.

Earthquakes in Canada
From 1899 to 1992

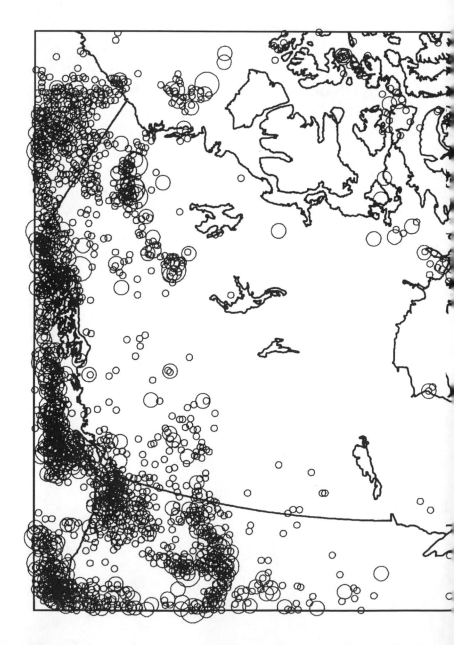

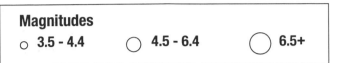

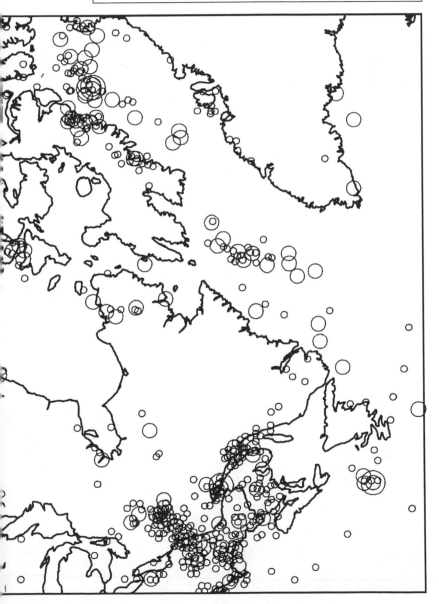

Chapter 9

HELPING EARTHQUAKE RESEARCHERS

O ver the centuries, hundreds of people have reported strange sounds preceding earthquakes, yet the phenomenon is still in question. The sighting of earthquake lights has been documented over the ages, but it was only recently that scientists set about trying to explain them. Time and again, volcanic eruptions have shaken the earth just before major quakes, but the global linking of volcanology with earthquake science remains almost as hotly argued today as it was when Sir William Hamilton first noted the seismic events that followed eruptions of Vesuvius.

Nor should it be lost on us that the theory of plate tectonics on which modern earthquake science is based did not become full-blown until 1968. We're talking about something rather new here, and we still have a long way to go in understanding earthquakes.

Can you help? Yes, you can, if you're an observant witness who is lucky enough to weather a good temblor. But for whom do you document it?

Interest in earthquake research is on the rise and scientists engaged in it are far-flung. Government leads the way in North America, with the following agencies among the leaders: U.S. Geological Survey, Geological Survey of Canada, National Science Foundation and the Federal Emergency Management Agency (both in Washington, D.C.), and the National Center for Earthquake Engineering Research at the State University of New York at Buffalo.

Also in the forefront are researchers at academic institutions such as the University of Colorado, Columbia University's Lamont-Doherty Geological Observatory, Cal Tech, University of California Berkeley with its Engineering Research Center, New York University, the Stanford University Earthquake Engineering Center, Boston College, Princeton, Texas Tech University (Lubbock), and the University of Alaska Fairbanks. Private industry, too, has good men and women in the field: architects, engineers, builders, psychologists, and planners who are working with peers in other nations to put this puzzle together.

It's rare, however, that personal accounts of earthquakes find their way into official reports and preparedness handbooks. For this material, we found ourselves relying almost entirely on the press, earthquake books, the unpublished notes of scientists, and reports compiled by observant laymen. So we suggest that you also consider us at Epicenter Press if you have earthquake experiences you would like to share. We'll pass your information on to an appropriate scientific agency, and we'll also consider it for the next edition of this book.

DISASTER PLANNING AIDS

- ☐ Conduct a Home Hazard Hunt.

- ☐ Reduce hazards in your home and work/ school environment.

- ☐ Establish an out-of-state contact.

- ☐ Assemble home, office, and car preparedness kits.

- ☐ Discuss your plans with family and friends.

- ☐ Locate your electric, gas, and water shut off valves and learn how to shut them off.

- ☐ Take a First Aid and/or a basic CPR course.

- ☐ Utilize the enclosed check lists.

- ☐ Rehearse your home and office/school response plan at least annually.

- ☐ Purchase at least one multi-purpose dry chemical fire extinguisher.

- ☐ Install smoke alarms and test on a regular schedule.

- ☐ Attend community preparedness programs and lectures. Stay informed and share information.

Hazard Hunt

Heavy Objects _____

walk through date _____

precautions taken by _____

Water Heater _____

precautions taken by _____

Gas Lines _____

precautions taken by _____

Electrical _____

precautions taken by _____

Water Valves and Lines _____

precautions taken by _____

Hazardous Materials _____

precautions taken by _____

Furniture Arrangements _____

precautions taken by _____

Kitchen Cabinets _____

precautions taken by _____

Gas Shut-off Wrench

DOs

1. DO keep out of reach of children.

2. DO let the gas company restore your service after you have shut if off.

3. DO seek the assistance of a plumber to repair gas pipe damage.

4. DO shut off valve if you smell gas and cannot locate leak.

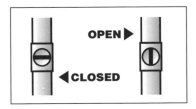

DON'Ts

1. DO NOT turn the gas back on yourself.

2. DO NOT shut off valve unless you smell gas.

3. DO NOT use matches or operate electrical switches if you suspect gas leakage.

4. DO NOT force valve when applying pressure.

5. DO NOT leave the wrench near gas meter.

6. DO NOT use wrench for any other purpose.

Draw a Plan of Your Home and Your Escape Plan Here

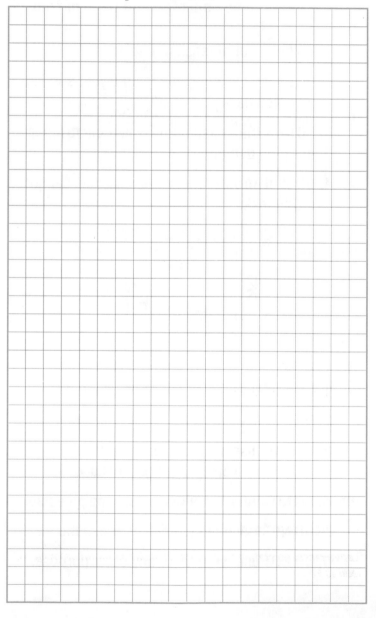

Evacuation Priority List

Survival items to be collected and hand carried.

❏ _____   ❏ _____

❏ _____   ❏ _____

❏ _____   ❏ _____

Other items in order of importance.

❏ _____   ❏ _____

❏ _____   ❏ _____

❏ _____   ❏ _____

Items to be removed if a car or truck is available and roads are accessible.

❏ _____   ❏ _____

❏ _____   ❏ _____

❏ _____   ❏ _____

Things to do if time permits, e.g., turning off utilities, locking doors and windows.

❏ _____   ❏ _____

❏ _____   ❏ _____

❏ _____   ❏ _____

Out-of-state family contact (___) _____

Our message location* _____

Leave message stating evacuation destination for others who will need to know.

SOURCES

Books

Association of Bay Area Governments. *The Liability of Businesses and Industries for Earthquake Hazards and Losses.* Oakland, Calif.: ABAG Publications, 1984. (Write to P.O. Box 2050, Oakland, CA 94604. Free for single copy. 415-464-7900.)

Bay Area Regional Earthquake Preparedness Project. *Earthquake Preparedness and Public Information Materials: An Annotated Bibliography.* Updated 1993. Earthquake Hazards Reduction, Series 8. Federal Emergency Management Agency and the California Seismic Safety Commission. (Write to BAREPP, MetroCenter, 101 Eighth Street, Suite 152, Oakland, CA 94607. 415-893-0818.)

Bloom, Roger. *In The Path of the Fault.* Tempe, Ariz.: Roger Bloom, 1984.

Bolt, Bruce. *Earthquakes.* Revised and expanded. New York: W. H. Freeman, 1993.

Boy Scouts of America. *Fieldbook: Boy Scouts of America.* Third edition. Irving, Texas: Boy Scouts of America, 1984.

Davis, Nancy Yaw. *The Effects of the 1964 Alaska Earthquake, Tsunami, and Resettlement on Two Koniag Eskimo Villages.* University of Washington PhD dissertation, 1971.

Dodgson, John H. *Earthquake and Structure.* New Jersey: Prentice-Hall, 1964.

Goodman, Jeffery. *We are the Earthquake Generation.* New York: Berkeley Publishing Co., 1978.

Kennett, Frances. *The Greatest Disasters of the 20th Century.* London, New York: Marshall Cavendish, 1975.

Mayse, Susan. *Earthquake: Surviving the Big One.* Edmonton, Alberta, Canada: Lone Pine Publishing, 1992.

McCann, John P. *How to Prepare for an Earthquake.* Updated 1990. New York: Insurance Information Institute. (Write to 110 William Street, New York, NY 10038. 212-669-9200.)

Merck, Sharp & Dohme. *Merck Manual of Diagnosis and Therapy.* 16th edition. Rahway, N.J.: Merck, Sharp & Dohme Research Laboratories, 1992.

Meyer, Larry L. *California Quake.* Nashville, Tenn.: Sherbourne Press, 1977.

U.S. Geological Survey

A lateral shifting of the ground beneath it effectively destroyed this structure in Alaska in 1964.

National Academy of Sciences. *The Great Alaska Earthquake of 1964; Human Ecology.* Washington, D.C.: U.S. Government Printing Office, 1970.

National Academy of Sciences. *Earthquake Safety: Activities for Children.* Washington, D.C.: Federal Emergency Management Agency. (Write to Earthquake Program, 5000 C Street SW, Room 625, Washington, D.C.)

National Science Teachers Association. *Earthquakes: A Teacher's Package for K-6.* Washington, D. C.: Federal Emergency Management Agency, 1988. (Write to National Science Teachers Association, 1742 Connecticut Avenue NW, Washington, D.C. 20009. 202-328-5800.)

National Security Council. *An Assessment of the Consequences and Preparations for a Catastrophic California Earthquake: Findings and Actions Taken.* Washington D.C.: National Security Council, 1980.

Steinbrugge, Karl V., with foreword by Charles U. Busch. *Earthquakes, Volcanos, and Tsunamis.* New York: Skandia America Group.

Sunset Special Report. *Getting Ready for a Big Quake.* Updated 1982. Menlo Park, Calif.: Sunset Books, Lane Publishing Co.

Verney, Peter. *The Earthquake Handbook.* New York, London: Paddington Press Ltd., 1979.

Yanev, Peter. *Peace of Mind in Earthquake Country.* Updated 1991. San Francisco: Chronicle Books.

Pamphlets and fact sheets

Coping with Children's Reactions to Earthquakes and Other Disasters. San Fernando Valley Child Guidance Clinic. Northridge, Calif.: Federal Emergency Management Agency and American Red Cross, 1983 and 1986. (Write to San Fernando Valley Child Guidance Clinic, 9650 Zelzah Avenue, Northridge, CA 91325.)

Earthquake Ready: Preparedness Planning for Schools. Oakland: Bay Area Regional Earthquake Preparedness Project, 1990. (Write to BAREPP, MetroCenter, 101 Eighth Street, Suite 152, Oakland, CA 94607. 415-893-0818.)

Earthquake Preparedness: Highrises—Mobile Homes. Ottawa: Emergency Preparedness Canada, 1990.

Earthquake Preparedness: People with Disabilities. Ottawa: Emergency Preparedness Canada, 1990.

Earthquake Preparedness Policy: Considerations for School Governing Boards. Office of the Governor (Office of Emergency Services); Sacramento, Calif.

Emergency Preparedness in Canada. Ottawa: Emergency Preparedness Canada, 1987.

The Emergency Survival Handbook. Los Angeles: The Red Cross, 1985. (Write to American Red Cross, Los Angeles Chapter, 2700 Wilshire Boulevard, Los Angeles 90057.)

The Family Earthquake Plan. San Francisco: Office of Emergency Services, undated.

Family Preparedness for Earthquakes and Other Emergences. Ottawa: Emergency Preparedness Canada, undated.

Learning To Live in Earthquake Country: Preparedness for People with Disabilities. Washington, D.C.: Federal Emergency Management Agency, 1985.

Learning to Live in Earthquake Country: Preparedness in Apartments and Mobile Homes. Washington, D. C.: Federal Emergency Management Agency, 1985.

The Next Big Earthquake in the Bay Area May Come Sooner Than You Think. Menlo Park, Calif.: United States Geological Survey, 1989. (Write to USGS, 345 Middlefield Road, Menlo Park, CA 94025.)

Prepare Now for an Earthquake in British Columbia. Vancouver: Insurance Bureau of Canada, 1991.

Tsunami! The Great Waves In Alaska. Wasilla, Alaska: Alaska Division of Emergency Services and Federal Emergency Management Agency, 1987. (Write to Alaska Division of Emergency Services, 3501 East Bogard Road, Wasilla, AK, 99687. 907-376-2337.)

What You Should Know Before You Hire a Contractor. Sacramento: Contractors State License Board. (Send a mailing label to CSLB, P.O. Box 26000, Sacramento, CA 95826.)

What We Have Done to Be Prepared for a Catastrophic Earthquake. Ottawa: Emergency Preparedness Canada, 1990.

Videotapes

Surviving the Big One: How to Prepare for a Major Earthquake. Los Angeles: Los Angeles City Fire Department and KCET Video, 1989. (Write to KCET Video, 4401 Sunset Boulevard, Los Angeles 90027.)

Quake of '89: A Video Chronicle. San Francisco: KRON-TV, 1989.

Principal agencies

Earthquake Engineering Research Institute, 6431 Fairmount Avenue, Suite 7, El Cerrito, CA 94530-3624. 415-525-3668.

Federal Emergency Management Agency, P.O. Box 70274, Washington, D.C. 20024. 202-646-2400.

U.S. Geological Survey, Earth Science Information Center, 345 Middlefield Road, Menlo Park, CA 94025. 415-329-4390.

A WORD
OF CAUTION

The United States Geological Survey printed the following disclaimer in its excellent survival guide titled "The Next Big Earthquake in the Bay Area May Come Sooner Than You Think":

> *This publication is meant to be instructional and to provide information that will help you understand and reduce the risk from earthquakes. The information in this publication is believed to be accurate at the time of publication. The agencies and individuals involved in the preparation, printing, and distribution of this publication assume no responsibility for any damage that arises from any action that is based on information found in this publication.*

The author and the publisher of this *Earthquake Survival Guide* invoke the same disclaimer.

ABOUT
THE AUTHOR

Born and raised in rural Maine, Lael
Morgan moved west to commute between
America's most earthquake-prone states
following graduation from Boston University's
College of Communication in 1959.

After serving as a reporter/photographer in
Alaska for the *Juneau Empire* and the *Fairbanks
Daily News-Miner,* she became a feature writer for
the *Los Angeles Times.* Following the San Fernando
quake she covered California's earthquake-planning
efforts and became fascinated with the subject.

Years later, after a long career as a freelance
writer for *National Geographic, Alaska Northwest
Publishing Co.,* and Doubleday, Morgan returned
to the far north to teach at the University of Alaska
Fairbanks.

She still pays earthquake insurance in two
states (Alaska and California) and claims to be
looking for a vacation home on a fault in Chile.

INDEX